WHY OUR MINDS WANDER

Understand the Science and Learn
How to Focus Your Thoughts

Arnaud Delorme

In association with

Published in 2023 by Welbeck Balance
An imprint of Welbeck Non-Fiction Limited
Part of Welbeck Publishing Group
Offices in: London – 20 Mortimer Street, London W1T 3JW &
Sydney – 205 Commonwealth Street, Surry Hills 2010
www.welbeckpublishing.com

A CIP catalogue record for this book is available from the British Library.

ISBN
978-1-80129-278-8

Typeset by Lapiz Digital Services
Printed in Great Britain by CPI Group (UK) Ltd, Croydon CR0 4YY

10 9 8 7 6 5 4 3 2 1

ABOUT THE AUTHOR

ARNAUD DELORME, Ph.D., has been studying human consciousness for the last 20 years. He is a Research Scientist at the Professor level at the University of California, San Diego, a CNRS Research Director in Toulouse, France, and a Research Scientist at the Institute of Noetic Sciences.

FOREWORD

Our stream of consciousness is curiously among both one of the most familiar and the most elusive aspects of our existence. Our inner world presents itself to us every day in endless ways. It often grabs our attention away from whatever task is at hand and takes us on unanticipated journeys, only to drop us off somewhere unexpected (e.g., miles past the road we meant to take). We routinely delight in flights of fancy and brood over dark thoughts that relentlessly return. Although present throughout our waking day, the nature of our inner thoughts is peculiarly difficult to grasp. As the eminent philosopher and psychologist William James (1890) observed over a century ago, "The attempt at introspective analysis... is in fact like seizing a spinning top to catch its motion, or trying to turn up the gas quickly enough to see the darkness".[1]

The meandering of the mind is not only elusive to introspection but also poses a serious challenge to scientific investigation. Indeed, for much of the last century, the internal stream of thought was considered off limits by behavioral scientists. In the early sixties, a bold handful of scientists began investigating the

---

1   James, W (1890). *The Principles of Psychology*, New York, Holt, p244.

stream of consciousness by exploring how the mind wanders away from the task at hand. However, the topic of "mind wandering" was largely ignored by mainstream researchers as too elusive to be rigorously investigated. Remarkably, little more than a decade ago if one were to peruse textbooks on psychology, or even cognitive psychology (a field explicitly dedicated to the mind), they would find no mention of this form of thought that we now recognize occupies up to 50% of our waking lives.

In recent years however, research on the topic of mind wandering has exploded. As Delorme details in this quite remarkable book, we now understand a great deal about how, when, and why the mind departs from the here and now. Like many things in life the concept of mind wandering defies a singular definition that captures the full range of circumstances to which it might apply. Nevertheless, when properly instructed, people can generally report when their attention is focused on the task at hand and when it has drifted to unrelated thoughts. Our capacity to make this very basic introspective judgment regarding the mind's focus (internal or external) provides a crucial leverage point for the scientific investigation of mind wandering. Researchers have now unequivocally documented that despite the challenges of introspection, individuals really can determine when they are mind wandering and when they are on task, as evidenced by the host of behavioral and neuroscientific differences that distinguish these states.

Drawing on this remarkable new body of research, Delorme takes us on a wonderous journey that illuminates a mental state with which we are all intimately familiar, and yet for which we have so much to learn. We gain the answers to such questions as: What exactly is mind wandering, and how do we recognize it? What is the nature of the brain when individuals report mind wandering? What impact does such drifting have on our performance and on our mood? What topic do our minds wander to? Under what circumstances are we most prone to mind wander and why? Is mind wandering problematic and, if so, what can we do to help to curb it? And if it is problematic, why do we do it so often? Might there be some hidden adaptive value to this routinely disdained state of mind?

In addition, to providing a rigorous yet highly accessible accounting of the science of mind wandering, Delorme introduces us to a host of valuable philosophical insights and lessons for how to live a happier and more fulfilling life. We are introduced to the perspective, routinely disregarded by the current mainstream neuroscientists but held by a notable minority of philosophers and scientists (such as William

James noted above), that the mind may reflect more than simply the machinations of a "brain machine." Rather than being the generator of consciousness, the brain may be more like a transducer, tapping into conscious fields much like the way a radio receives electromagnetic waves. While still recognizing the important role that the brain plays in directing consciousness (a broken radio can seriously distort its signals), this perspective opens the possibility that the meanderings of the mind may be impacted by factors outside the normal scope of mainstream science.

Although Delorme acknowledges his atypical (but in my view, quite refreshing) take on the relationship between consciousness and the brain, his book does not require that one adhere to this perspective in order to gain deep wisdom. Throughout the book, we learn a host of powerful techniques and activities that enable us both to become more familiar with our wandering minds, and to learn how to effectively manage them. We are introduced to the power of meditation, and basic techniques for how to avoid letting our thoughts get the better of us (hint: just because a thought arises does not mean one has to endorse it). We learn some of the basic principles of cognitive restructuring and how to avoid going down the rabbit hole of catastrophizing. We discover valuable methods for working with the body, diagraming our thoughts, and enhancing meta-awareness (gaining a real-time understanding of the current content of one's mind). We are also introduced to a host of lesser-known approaches, including: the happiness technique, which helps us to discover fallacies we may hold about what makes us happy, and the tapping technique, which entails distracting the mind from challenging thoughts by simply tapping the body with two fingers.

After finishing this book, readers will have a new-found understanding of the internal terrain that they visit every day: recognizing familiar landmarks but with a new found appreciation of how to navigate them. With the perspective and tools that this book provides, readers will be equipped to both effectively manage their mind wandering, and to discover for themselves the potentially unique ways in which they flow down the stream of consciousness.

Jonathan Schooler
Distinguished Professor of Psychological and Brain Sciences, UCSB
Director, Center for Mindfulness and Human Potential

CONTENTS

INTRODUCTION

Marie is reading her book when her mind drifts into thinking about her recent vacation in Hawaii. As she pictures herself back on the beach, she is stuck in her memories and does not realize that she has stopped reading, although her eyes are still moving on the page. After some time, she realizes she has lost track of the words in the book. She will need to reread a few pages as she had been completely distracted. Marie was daydreaming, or in technical terms, "mind wandering." Her mind wandered away from reading, even though she wanted to read. Why did it happen? Instead of daydreaming while reading, why didn't she put the book down, close her eyes, and enjoy remembering her vacation?

John is stressed. He has read on social media that meditation might help him relax so he decided to try a simple breath meditation every day for 15 minutes. On day one, John sets a timer on his phone for 15 minutes and gets ready to meditate. At first, he is entirely focused on his sensations. He can feel the air moving in and out with his breath and his chest's movements. Focused, he is observing every sensation. Suddenly, he experiences a fleeting thought, "This is so nice and relaxing, I love meditating and emptying my mind." The thought goes on, "This meditation is going to relieve me of all my stress. I just cannot stand the deadlines and pressure anymore. They are not paying me enough . . ." His mind goes on and on. At this point, John has lost track of his meditation because he is thinking about his worries and work stress, his breath has become shallow and his body has contracted. He remembers, "Oh gosh, I am not focused on my breath anymore. Let me go back to my meditation." He goes back to his meditation, and a few seconds later a fresh stream of thought pops up, "I am such a loser for not being able to focus on my breath. This reminds me of when . . ." Again, he has lost track of his breath. John comes in and out of his meditation until his phone's timer rings. Why couldn't John focus his mind while meditating? Is he mentally impaired, or is it part of the human condition?

Bobby is a 15-year-old boy. He loves math and is good at it, when he focuses. The problem is that he does not seem to want to concentrate. Bobby is obsessed with video games and this is apparently the only thing he can focus on. When it is time to do his homework, he can do the short math problems, but the long problems are too much. He tells himself that they are boring, but really his mind cannot hold the thought long enough to solve the problem. He reads the problem and starts thinking about a potential solution, but then a random thought disturbs the flow of his thinking. It could be anything from "What's for dinner?" to "I need to finish the next level on my new video game" or "This is so boring, I cannot do this." Sometimes, he just feels antsy and has to stand up. After these episodes, he has forgotten the math problem and needs to restart, but when he does, the same thing happens again. Bobby's capacity for focusing is impaired because random thoughts pop into his mind. It is not something he can control consciously.

Jane is an engaged person. She has political views and likes to challenge people on online forums. She feels she is right. In fact, she knows she is right. The presidential election is approaching and she idealizes her preferred candidate and demonizes the one from the other party. Her candidate loses the election and she has a hard time believing it. In fact, she cannot believe that the other candidate, who is obviously evil, has won. The thought permeates her mind on all occasions. She does not want to have these thoughts anymore as they make her miserable. Yet, she cannot help it. This is an extreme example of mind wandering, called rumination, where the same thought is rehearsed over and over and forces itself into people's minds. How is it possible that Jane, who actively does not want to have these thoughts, cannot help having them?

Questions this book will answer

I am a research scientist of neuroscience at the University of California San Diego and a scientist at the Institute of Noetic Sciences (IONS) studying how the brain processes information to give rise to our experiences. This book is about what happens when our minds wander. It examines why, when we are focused on a task, we cannot stop our minds from daydreaming or wandering away from the task at hand, what is going on in our heads when this happens, and what we can do to gain greater control over our minds.

Mind wandering also taps into the big picture of where our thoughts come from and whether we are our thoughts. Are we more than our thoughts? How can we tame our thoughts? For example, when you are trying to read this book, are you able to ignore distracting elements in your environment? If you are trying to complete a monotonous DIY (Do It Yourself) task in your home, can you focus and ignore the activity around you, or do you get distracted easily? What can we do when excessive mind wandering prevents us from living a normal life, from focusing when we need to concentrate, or from reaching our full potential?

Our thoughts are central to who we are but the science and mechanics of exactly how our minds work is only just starting to reveal its mysteries.

In **Part 1: What Is Mind Wandering?** we look at how scientists study the wandering mind to help us understand where mind wandering comes from, why it can be problematic, and what we can do about that.

Part 2: Mind Wandering and You provides a unique window into the human mind, looking at the different types of thoughts that make up mind wandering, and what research can tell us about them. We will also delve back into the history of the study of mind wandering, and how scientists have developed techniques to uncover how people's thoughts are influenced by their mood, attention and personality. We will also discuss how involuntary thinking interacts with different areas of our lives. Do you spontaneously have negative thoughts when you feel down? How does the content of involuntary thoughts affect your mood?

Moving beyond psychology into neuroscience, brain-imaging techniques can determine what happens in the brain when we think about something. As more becomes known, understanding the neurological mechanisms underlying our thinking will help us to devise new methods to curb problematic thinking patterns. Ultimately, this is the key to being able to direct our thinking to help us live better lives.

Part 3: Taming Your Mind brings together an array of methods to influence the content and frequency of our mind wandering, with the goal of improving our quality of life and controlling unwanted pervasive thoughts. Learning about why these methods must be indirect and why we cannot just intellectually convince ourselves to stop having some thoughts, I present techniques to discipline the mind and curb mind wandering to help you achieve greater levels of happiness. Powerful dialectic and meditation practices allow us to reduce the grasp of damaging thoughts, leading them to eventually disappear. The secret lies in being able to view our thoughts from a new angle and the ability to retrain our minds by working with our body's sensations.

By the end of this book, you will have a clearer picture of how your mind works and an array of practical methods to become a better version of yourself.

PART 1

WHAT IS MIND WANDERING?

CHAPTER 1

THE DIFFERENT TYPES OF THOUGHTS

Before we start, what even is mind wandering? If the concept of mind wandering is confusing to you, you might be surprised to find out that it is also still unclear to scientists and researchers in the field.

Researchers assign different meanings to what it is to mind wander. Some researchers include intentional thinking within mind wandering, while others only consider thoughts to be mind wandering when they are unintentional. There is no "right" or "wrong" answer.

Types of mind wandering

What constitutes mind wandering will differ based on which research team you are speaking to. In the diagram on the next page, we show different types of thoughts and how intentional and varied they are. Let's consider these key types of thought.

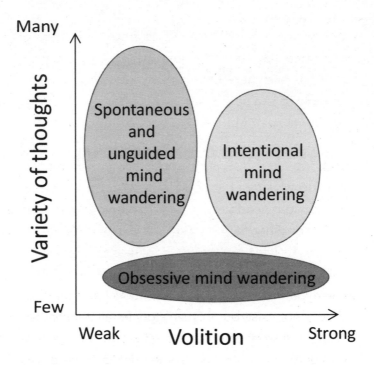

Intentional thoughts

Intentional thoughts are the goal-directed thoughts we have when we intend to solve a problem and direct our minds to do so. The content of the thought is directly controlled by our will. You could be thinking about what you will buy at the grocery store, or, if you have a gardening project, willfully thinking about the different steps is an intentional thought. These thoughts are usually associated with achieving a specific goal.

Why might some scientists consider intentional thoughts as mind wandering? Thoughts are often not black or white when it comes to being voluntary or involuntary. For example, imagine you are washing the dishes and then start thinking about your plan for tomorrow. It is a spontaneous thought, or can you be sure you willed it? Maybe it just popped into your mind – but you are washing dishes so you started thinking about food by association, and then on to grocery shopping which you happen to have to do tomorrow. One thought leads to another, and you may start thinking of other things you need to do tomorrow. It's not clear if this is voluntary or triggered by you washing dishes. Some scientists even argue that there is not a single type of thought that is intentional and that free will does not exist, but this is a story for a later chapter.

When we work with our thoughts, we will see how intentional thoughts can sometimes sabotage our life and that, even though we think we control them, sometimes they control us.

Unintentional, spontaneous thoughts and daydreams

While there is debate among scientists about intentional thoughts as mind wandering, most agree that unintentional spontaneous thoughts do count as mind wandering. Spontaneous thoughts are the type of thought Marie experienced while reading a book and those John had while meditating. These are typical mind-wandering thoughts – involuntary and, as hard as we try, we are usually unable to prevent them from occurring. These thoughts might have their use (an interesting idea may pop up) as we will see later.

Often spontaneous thoughts are self-centered and participate in the "story of me." The degree of spontaneity might vary. When performing a boring task, we might welcome a spontaneous thought, and believe we initiated the thought. When trying to focus, particularly during meditation, we may judge the same thought as distracting. The common characteristic of these thoughts is that they arise spontaneously, and we usually have not willed them into existence.

Spontaneous thoughts are often unconscious. If you are planning to meditate, whether it is a guided meditation or self-paced meditation, you intend to focus on the meditation. Yet, your mind will invariably go to other places. Even though you remain conscious of your thoughts at all times when your mind starts drifting, you forget you were supposed to be meditating. We are aware we are thinking, but we are unaware when our mind is wandering. Once we realize we have these thoughts, we are surprised by their presence and come back to the task at hand. Spontaneous thoughts share some characteristics with dreaming. As we dream, we are aware we are thinking but not aware we are dreaming. Researchers usually consider that mind wandering and dreaming lack the same elements of self-awareness.

However, not all spontaneous thoughts are distracting, and I want to bring your attention to a subtype of spontaneous thoughts: unintentional creative thoughts. These thoughts occur when we do not strongly direct our mind and we observe it wandering to gather ideas on a topic. For example, when writing a letter or message to a friend, we may let our minds wander about what to write. Or when drawing, we often follow our inspiration. Creative thought has to do with intuition, while goal-directed thoughts usually rely more on reasoning. Although, again, it is not a sharp divide, and some thoughts will borrow from both.

Daydreaming may also be considered an unintentional spontaneous thought. Daydreaming occurs when we imagine ourselves in pleasant scenarios – perhaps imagining how life would be if we had unlimited resources: the house we would buy, the travel we would take, and the gifts we would give to our friends and relatives. While some see daydreaming as a synonym for mind wandering, daydreaming is more of a subtype of mind wandering, specifically referring to positive imagination.

We will look into all types of spontaneous thoughts in detail, especially the ones that make us unhappy. We will look for their roots and also assess how we may curb them. Although we do not have direct control over these types of thoughts, there are mental exercises we can practice to positively influence them.

Stimulus-dependent and stimulus-independent thoughts

These are thoughts triggered by external events. If I hear a bird chirping outside, this could remind me of when I was young and listening to the birds in the garden of my grandparents' house, and the good times I had there. It is called a stimulus-dependent thought. If I did not hear the birds, I would probably not have had that thought. Some researchers argue that stimulus-dependent thoughts do not qualify as mind wandering, while others think they do. To determine if someone has stimulus-dependent thoughts, researchers ask them to perform a task, and interrupt them asking them what they think, and the context of what they are thinking. If their current thought was triggered by an external event (such as a dog barking outside), then it is a stimulus-dependent thought.

Again, the distinction between stimulus-dependent and stimulus-independent thought is hardly clear cut. What if my stomach rumbles, and I start thinking about food? Is it a stimulus-dependent thought or not? We are also seldom in a floating tank with headphones, completely isolated from our environment. In these conditions, it is difficult to determine what stimulus-independent thoughts are, as they could all be triggered by minute changes in our environment.

Stimulus-dependent thoughts are important as external events often trigger some cognitive response. Say your partner or roommate never empties the dishwasher and you feel resentful about that. Every time you notice the dishwasher running, this thought might cross your mind. Added to other similar resentful thoughts, this may make you unhappy. We will discuss ways to deal with these triggered thoughts and how to retrain our minds.

Task-independent thoughts

People usually start mind wandering when performing a common or mundane task – like washing the dishes or sweeping up leaves. Most people are thinking of something and often it has something to do with the activity they are engaged in. If you spontaneously start thinking about dishes while washing the dishes or driving while driving, this is a task-related thought. Some researchers would say it is mind wandering while others that you are simply thinking about what you are doing. Again, differentiating between the two is not simple. Imagine you are driving and start thinking about doing the dishes because when you were young you went out with your parents in an RV with your dad washing the dishes while your mum was driving. So, the thought of washing the dishes would actually be a driving-related thought for you. Do you see the dilemma?

Task-independent thoughts are important. We will see that when we have too many of them, we may fail to concentrate. We will discuss how to deal with these distracting thoughts, and how to decrease their occurrence.

Unguided thoughts

Intentional and unintentional thoughts refers to the thought onset. By contrast, guided or unguided thoughts refers to your voluntary control of the stream of thoughts as they happen. For example, if you are thinking about your plan to go grocery shopping tomorrow then you decide to think about the rest of your plan for tomorrow, does the thought occur spontaneously? Are you guiding your stream of thoughts, or is it just "happening", with you as a conscious but passive observer? Some researchers believe thoughts only qualify as mind wandering if they are unguided.

Typical unguided thoughts are meandering thoughts that jump from topic to topic without a specific logic. Say you are thinking about your plan tomorrow, then start thinking about your childhood, then about sports on TV. These thoughts are not related, and the logic behind their sequence eludes you.

Unguided thoughts can also occur during a dream. In a dream, we may have thoughts and later become conscious of them, but we do not consciously control them. These thoughts are at the lowest level of volition. We often do not even remember them. They are deconvoluted, jumping from one subject to another unrelated one, and obeying different rules not subject to mental censorship like waking thoughts. When dreaming, believing in being a Nazi or a god could be completely acceptable, as are actions that violate the laws of time and space, like, for example, talking to a deceased relative or flying. We may even believe we are someone else.

Unguided thoughts are the type of thoughts that occur during dreams, and as we will see later, when they occur during waking hours they might correspond to microsleep, allowing our mind to rest. We will see why becoming aware of them and our automatic thinking patterns is critical to increasing well-being.

Compulsive, ruminative thoughts

When thoughts are always about the same topic and they repeat constantly, we call them obsessive or compulsive. In mental health literature, these are also called ruminations. Ruminations are a type of involuntary thought between goal-directed thoughts and spontaneous thoughts. To determine if a thought is obsessive, researchers have participants do repetitive tasks and interrupt them asking them what they think. If they always think about the same thing, and if the thought is intrusive and makes them unhappy, it is a ruminative thought. If we have ruminative thoughts, we are often aware we have them. We earlier gave the example of Jane's rumination about her political belief. She cannot stop thinking about specific people and behavior. She thinks about it several hours a day, and this makes her unhappy.

When we focus on controlling our thoughts, it is usually this type of thought we are referring to. Maybe you have picked this book because you have ruminative thoughts. We will see in a later section how we can deal with these thoughts.

* * *

These classifications are useful for capturing some essential characteristics of mind wandering. While researchers disagree on the definition, they agree on the underlying phenomena. In the history of mind-wandering research, we will see that mind wandering is a newly defined term, which is why some researchers take the liberty to include more or less phenomenal experiences, and create their own definition.

In the rest of this book, I will use a broad definition of mind wandering that encompasses all the types of thoughts described above. In our daily life, we rarely differentiate between different types of spontaneous and nonspontaneous thoughts. We listen to our "mental radio," which sometimes plays thoughts relevant to the task we are doing, and occasionally irrelevant thoughts. Sometimes it also plays meta-thoughts – thoughts about thoughts – which try to influence other thoughts, such as "I should not be thinking this" or "I am not focused enough on what I am doing."

When trying to deal with our thoughts, for both spontaneous and intentional thoughts, what is important is the power they have for us to believe them. The more we believe a thought, the harder it is to let it go. We will do several exercises in this book to become aware of our mind wandering. The first exercise below helps you to start to understand what is going on when your mind wanders.

EXERCISE: LISTENING TO YOUR THOUGHTS

I remember the first exercise my meditation teacher gave me. It was not a meditation exercise per se. Instead, the task was to sit at a desk, listen to my thoughts, and write them down. It could be daily events, philosophical reflections or solving problems; it was eye opening. Let's try:

- Take a blank piece of paper and a pen. Then sit in a quiet place where you know you will not be disturbed. Set a timer on your smartphone for 10 minutes.
- Write down all the thoughts that cross your mind during these 10 minutes, writing each thought on a new line.
- If your thoughts come faster than you can write, just write what you can.
- At the end of the 10 minutes, try to classify each thought into one of the categories above: intentional or spontaneous, task-oriented or not, directed or not.

Below is an example of me doing the exercise (after removing some repetitive thoughts). I have also labeled the type of thought involved.

- What am I going to talk about? Oh wait, this is about my personal thoughts not about sharing with others (task-related thought).
- Unstructured fragments of thoughts about future meetings. Unclear (spontaneous thoughts; unguided thoughts).
- Something is vibrating on my desk as I write (stimulus-dependent thought).
- The sun is nice outside and I can hear a motorcycle. It is weird to have a motorcycle on that road (stimulus-dependent thought).
- OK, a few more minutes to go. What I write might be boring for others to read. Am I going to include that in the book? What will people think if I do? (task-related thought).
- But it is good that I have different types of thoughts (task-related thought).
- How am I going to organize all of this if I include it in the book? (intentional thought, goal-oriented thought).
- No thought crossed my mind for 10 seconds, that is strange (task-related thought).
- I am rubbing my eyes but I have just handled some used batteries. Is that safe? (stimulus-dependent thought).

Often when doing this exercise, your mind starts racing, and you can barely write fast enough. As in my attempt above, there are often a lot of task-related thoughts: you may ask yourself why you are doing the exercise, tell yourself it is fun or boring, or worry you are not doing it right (by the way, you cannot do it wrong if you write everything that crosses your mind). We also see a number of stimulus-dependent thoughts triggered by the environment. There were some spontaneous thoughts, which were vague and not fully formed about a future meeting, and an intentional goal-directed thought where I was already planning how I would organize this section.

The goal of the exercise was to show that the mind is always active, and also all over the place. This is the nature of the mind and does not mean that there is something wrong. As strange as it seems, when we are engaged in daily activities, we often fail to realize our mind is always busy. We live through our thoughts but we only recognize how active our mind is when asked to monitor it. Now you have seen the different types of mind-wandering thoughts going through your own head, perhaps it's time to learn more about where those thoughts come from.

CHAPTER 2

THOUGHTS ABOUT OURSELVES

The thoughts and beliefs we have about ourselves are the most significant, as they define us, and these are the thoughts that come back to us over and over again when we are mind wandering. Our thoughts about ourselves can be categorized into different types, such as self-critical thoughts, self-affirming thoughts, self-doubting thoughts, self-judgmental thoughts, self-comparing thoughts and self-reflective thoughts. Self-critical thoughts involve negative beliefs about our abilities, such as "I'm not smart enough," "I always mess things up," or "I'm not attractive enough." Self-affirming thoughts are positive and empowering, such as "I am capable," "I am strong," or "I am worthy." Self-doubting thoughts make us question ourselves. "Can I really do this?" or "Am I making the right choice?" Self-judgmental thoughts lead us to criticize our own actions or behaviors, such as "I shouldn't have said that," or "I'm such a failure." Self-comparing thoughts involve comparing ourselves to others, such as "Why can't I be as successful as them?" or "I wish I had their life." Lastly, self-reflective thoughts help us understand our thoughts, feelings and behaviors, such as "Why am I feeling this way?" or "What can I do differently next time?"

By being aware of the types of thoughts we have about ourselves, we can better manage our mental health and well-being, as negative thoughts can lead to low self-esteem and depression, while positive thoughts can boost our confidence and self-worth. The practices in this book will hopefully guide you to that place beyond thoughts and belief, a place where people report experiencing more peace, happiness and capacity to focus. Thoughts may gradually stop running your life, and you come to rest in what you really are.

EXERCISE: "WHO AM I?"

There is a revelatory exercise in psychology where you challenge people to define themselves by asking: "Who am I?" You might start by giving your name, but then you have to continue with further elaboration. Let's try:

1. Take a blank piece of paper and a pen and sit in a quiet place where you know you

will not be disturbed. Set a timer on your smartphone for five minutes.

2. Now write down who you think you are. There is no need to make sentences – you could write a single word. In my case, I would probably start with "a father," then "a neuroscientist," for example.

3. After you write something, imagine someone asking you, "OK, please tell me more."

4. It is OK to pause for 30 seconds or more to just think deeper about who you are. When you are ready, write as many definitions of yourself as you feel necessary.

5. When done, come back here.

First you write down definitions of yourself – such as your gender, nationality, marital status, your job, your education and so on. It starts becoming interesting when you keep pushing: "Tell me more. Who are you?" At this point, you might describe some personality traits: "I'm shy" or "I'm an extrovert." Eventually you might end up on more existential concepts: a human being or a soul depending on your religious or spiritual orientation.

Let me try this exercise on myself. This will also be an opportunity to introduce myself and how I came to study mind wandering.

"I am a kid who grew up to be an adult."

My adventures with mind wandering actually started in 7th grade when I was about 12 years old. Before that, I wanted to become a firefighter, but one day I was in the courtyard in my suburban Paris school and was struck by the random thought: "I want to understand why I am here." I remember this episode clearly, although, over the years I have probably romanticized it. It was an existential question and to me, with my Western upbringing, that meant understanding the brain and becoming a brain scientist. Had I been born in the 12th century, asking such questions might have led me to become a monk. It stuck, probably because the answer to these existential questions – assuming there are some – is not straightforward.

"I am a scientist."

Fast forward 15 years and I finished my Neuroscience Ph.D. in Toulouse, France and moved to the Salk Institute in San Diego to pursue my studies. I was surrounded by some of the most talented scientists studying the human mind, yet realized we know so little. I remember feeling intimidated at a lab meeting presenting in front of my peers, including Francis Crick, the Nobel Prize winner for his discovery of the DNA who was interested in consciousness research, and Bernard Baars, a well-known philosopher of consciousness. Discussing models of consciousness on this occasion and others, I realized that science would never be able to answer the question of *why* we are here, because this is not what science is about. Science is about telling us *how* things work, not *why* things are the way they are.

"I have thoughts."

Secondly, I am not only conscious, but I have thoughts and feelings. This sounds like a philosophical statement, but it is in fact very practical. These can be thoughts of being hungry, thoughts about future work or leisure plans, or thoughts of wanting to talk to a friend. Thoughts are usually associated with feelings: I can feel excited, sad, frustrated, content, etc. They also come with different levels of intensity. I have thoughts all the time, except when I sleep. My thoughts and feelings come and go, and sometimes it does not feel like I control them. An unexpected raise at work means I am elated, then a letter may arrive from the tax administration and my mood and thoughts become dark. Like the characters at the beginning of this chapter, sometimes I do not want to have thoughts but they still occur.

However, through the practice of meditation and other techniques we talk about later in this book, we may come to understand that we are not our thoughts. Thoughts come and go, and we are not what comes and goes. We are the background over which this happens. Thoughts, irrespective of what they are, remain mental constructs.

"I am conscious."

I am, like you, a conscious and alive being. By "conscious," I do not mean moral, but aware. I am a person who perceives; I can watch a beautiful sunset and be in awe. What does that mean? You are alive so you can understand that concept, but it would be extremely hard, if not impossible, to explain this to an animal or a computer. We know it intuitively. Hundreds of philosophical books and papers have been written on the subjective feeling of being alive and conscious. There are strong debates in the scientific community as to whether our subjective sense of self can be reduced to a mechanistic process. In other words, are the thoughts in our mind more than the activity of the neurons in our brain? We will touch on this again later.

Working on the hard problem of consciousness

In science, we have a reductionist approach where the scientific edifice is like a pyramid compartmentalized in different levels and we each work at our specialized level within the pyramid. In this view, matter is made of atoms, which make molecules, which make proteins, which make cells, which make the human body, which makes consciousness. Say a scientist named John works on brain cells. John does not question why molecules and proteins within the neurons interact the way they do or worry about higher-level cognition. He leaves that to other scientists. Instead, he focuses on how neurons store information and communicate with each other, leaving the ultimate mystery of why things are the way they are to other levels in the pyramid.

But this is right. Particle physicists at the bottom of the pyramid have no explanation as to why particles exist or behave the way they do. In fact, a scientific motto in quantum mechanics, known as the Copenhagen interpretation, is "Shut up and calculate." I am one of the researchers at the top of the pyramid, trying to make sense of the lower levels and how human consciousness emerges from inert matter. This has been termed the "hard problem of consciousness." Why? Because, intuitively, a mechanical machine like the brain should not be able to give rise to subjectivity.

Imagine we have a special camera capable of imaging all the neurons in the brain at very high speed, and we are able to see the sequence of neurons becoming active in the brain. Over a picosecond, which is a very short time (there are 1,000,000,000,000 picoseconds in one second) only one event may happen – a neuron stimulating another one for example. There is no subjectivity there, only biochemical processes, and we might be able to see every single one of them at that speed. I cannot really grab a magic wand and declare that if I speed things up, subjectivity magically appears from these biochemical processes. A small community of scientists to which I belong thinks that inert matter cannot give rise to conscious experience. This is a highly debated question which we will come back to later.

Realizing we are not our thoughts

Based on this awareness that science would tell me how things work, but not why they are the way they are, I became interested in other approaches to studying the mind, in particular meditation. In meditation, you watch what your mind is doing. You study your mind from the inside. I started with the book *Meditation for Dummies*, then joined a Zen center. The first thing I realized in meditation was that I was not my thoughts, especially my emotional thoughts.

I could believe a thought intensely, perhaps a grudge about a friend, then later during the day become indifferent about that thought. Sometimes this change of perspective happened in seconds, and when it did, I realized that the person holding that grudge a couple of seconds before was not me. It was just a thought that had captured my full attention and that I had consciously or unconsciously chosen to believe. The second thing I realized was that when meditating I spent most of my time mind wandering, even after several week-long meditation retreats. Because mind wandering was not being looked at by the neuroscientific community at the time, I decided to study this topic in my academic research and better understand this process.

A turn in my career happened when I was at a meeting with my professors and peers discussing a presentation on computational models of the mind in 2002. The presentation was a complex model of how the mind works, with concepts inspired by the way computers work. I watched the presentation and then, in front of all my colleagues, asked this question: "Say a book exists, describing all the computational processes in the brain and how they theoretically give rise to consciousness. It is a very big book with thousands of pages, and you can barely lift it off the table." I gestured over how to handle this huge book on the table. "What does it tell you about yourself?" I wanted to explore who we are as thinking human beings. My point was that rational thinking is just that – "thinking," not being. My peers and professors looked at me puzzled, not understanding my question. They could not imagine that solving the problem of consciousness would involve anything other than mechanistic processes. At this point, I knew I had to step out of mainstream academia to explore fringe questions about consciousness.

I began collaborating with the Institute of Noetic Sciences (IONS), an institute created by astronaut Edgar Mitchell and dedicated to studying fringe questions in science. Edgar Mitchell is one of the 12 humans to have walked on the moon. On his way back to Earth, he had an epiphany. In his own words (Sington, 2007):

"The biggest joy was on the way home. In my cockpit window, every two minutes: The Earth, the Moon, the Sun, and the whole 360-degree panorama of the heavens. And that was a powerful, overwhelming experience. And suddenly I realized that the molecules of my body, and the molecules of the spacecraft, the molecules in the body of my partners, were prototyped, manufactured in some ancient generation of stars.

*And that was an overwhelming sense of oneness,
of connectedness; it wasn't 'Them and Us,' it was
'That's me!', that's all of it, it's... it's one thing.
And it was accompanied by an ecstasy, a sense of
'Oh my God, wow, yes,' an insight, an epiphany."*

Mitchell created the Institute of Noetic Sciences in 1973 and raised funds to address the reality of these experiences – that consciousness is a property of nature, not an illusion or a mechanistic brain process. If one can show that telepathy is real, for example, then this is a step forward to show that consciousness is a fundamental property of nature. As we will see in this book, it also has consequences regarding where thoughts come from and how mind wandering arises. Let's learn more about the study of mind wandering . . .

CHAPTER 3

THE STUDY OF THE WANDERING MIND

There is a discrepancy between our perception and what neuroscience tells us is happening inside our brains. Neuroscientists argue that visual consciousness happens at a rate of about 10 times per second (VanRullen, 2016). You can imagine our visual cortex scanning the environment multiple times per second. This is not our experience, of course; our perception of time and our environment is continuous. This is why Professor William James, a renowned Harvard scientist at the beginning of the 20th century, argued that we must use introspection to study consciousness (James, 1890). Introspection is the process of examining your own (and others') thoughts and perceptions, and in this chapter we will learn how scientists use introspection to find out what is going on in our minds. Later, we present what neuroscience has to say and how we reconcile the two approaches and, in turn, how this impacts how and why our minds wander.

The "monkey mind"

Let's look at the origin of introspection in the study of spontaneous thoughts. The mention of spontaneous thoughts can be traced back to 4th-century ancient Chinese texts and later Buddhist texts of Japanese origins. These texts refer to the "monkey mind," where the meditator constantly loses their focus due to spontaneous thoughts. The human mind is compared to a monkey jumping from branch to branch – our mind cannot stay still and thoughts come and go constantly. The Christian tradition has also touched on this, describing how thought arises during prayer in middle-age texts and how to purify the mind from such thoughts. For example, in the mystic 14th-century Catholic text, *The Cloud of Unknowing*, we can read in Chapter 7: "How a man shall have him in this work against all thoughts, and specially against all those that arise of his own curiosity, of cunning, and of natural wit." (Anonymous, 2018). This text emphasizes the wariness of the thoughts that arise spontaneously and randomly.

The term monkey mind is not used in scientific circles. Instead, neuroscientist Jean-Philippe Lachaux has coined the term "mental radio" to describe this phenomenon: the radio is always on. Like a radio, our mind constantly receives and embeds "thoughts" about various topics throughout the day: "your mind wanders because your brain whispers all the time." Sometimes, we can be neutral about the thought content and simply listen to our thoughts as we would listen to the radio but at other times this is not possible. For example, the mental radio might be especially active when you are listening to a politician you do not like talking on the TV. In this instance, our mind might spontaneously comment as to why every statement spoken is wrong.

It was not until the late 1800s that scientists and psychologists became interested in studying this phenomenon. The study of introspection – asking people what they think – had its golden age in the late 1800s with Professor William James as the beacon for these movements. He wrote: "Consciousness is in constant change. (. . .) Now we are seeing, now hearing; now reasoning, now willing; now recollecting, now expecting; now loving, now hating; and in a hundred other ways we know our minds to be alternately engaged."

This coincided with Sigmund Freud who also wrote extensively about daydreaming. (In this era, mind wandering and daydreaming were not considered separately.) For Freud, daydreaming was a negative phenomenon, making us enact our unfulfilled fantasies to respond to a deprivation state (Freud, 1908, 1962).

These early efforts were not structured research, but rather essays based on interviews with people or patients. We call them qualitative approaches. It was not until the second half of the 1900s that psychologists designed quantitative approaches and questionnaires to study mind wandering. The 1970 *Imaginal Processes Inventory* was the first known scale in psychology to address mind wandering, with questions such as "I am the kind of person whose thoughts often wander?" It was innovative, because with this type of questionnaire, it is possible to obtain yes/no answers from thousands of individuals and to finally have some statistics.

For most of the 20th century, studying the mind and consciousness – and as a consequence mind wandering – was a taboo in science. The behaviorist movement argued that studying such questions was unscientific. Behaviorism considers the brain a "black box" – processing input and delivering output. Thoughts and unconscious processes leading to them were considered irrelevant, and they argued that only people's behavior, which can be observed, should be used in the study of human cognition. This was strongly influenced by the creation of computers and the idea that the brain was a sophisticated computer. With a computer program, you feed it input, and it outputs results based on its programming. By studying the input and output, you may infer how the program works.

Cybernetics: the brain and its environment

Then, in the 1990s, came the idea that organisms were not independent of their environment and formed a complex system with it. Considering the brain as a black box to which we provide input and study output is flawed, because our environment and our interaction with others is part of our cognition and not independent of it. The study of cybernetics – the science of communication between living things but also machines – paved the way for modern cognitive neuroscience. The study of introspection became popular again.

Take an extreme case of someone attacking another person, and trying to explain the attacker's behavior. It is not enough to enumerate the facts of what happened to the attacker. Another person having gone through the same sequence of events might not have the same behavior. One must also consider the personality, potential mental disorders, and also the type of relationship the attacker entertained with their victim. To better understand this person's reasons for attacking someone else, it would be necessary to interview them and run psychological tests. For example, say you observe someone from a partially blocked view pushing someone else in the street. You wouldn't know who the true aggressor is, who is in self-defense, or which person is truly being aggressive.

It could also be that the person pushing is trying to save the other person from being run over by a vehicle outside your field of view. Intentions matter!

Yoga and meditation

Another important societal phenomenon that pushed scientists toward the study of mind wandering was the interest of the general public and scientists in meditation and Yoga. In 1991, the Dalai Lama founded the Mind and Life Institute. To promote the scientific studies of meditation, this institute brought together contemplative Buddhists and gave small grants to young scientists to study meditation – I was fortunate to be awarded one of them on the study of mind wandering during meditation in 2004 – we will talk about that study later. In July 2001, the *Times Magazine* front cover was about the science of Yoga.

The study of meditation in neuroscience boomed in the early 2010s and, along with the study of spontaneous thoughts during meditation, we are still riding that wave. Professor Jonathan Schooler at the University of Santa Barbara, an esteemed colleague, friend and the author of the foreword of this book, was one of the first to bring mind wandering to modern psychology and neuroscience with landmark studies on how mind wandering varies across populations or environments. According to the article database of the National

Institute of Health in the US, there have been more than 1,500 scientific journal articles on mind wandering published in the last two decades, and we will review key articles and studies over the coming chapters to see what they can teach us about our wandering minds.

The default mode network

Our account of the historical development of the study of mind wandering would not be complete without mentioning another independent line of research that appeared in the beginning of the 21st century in neuroscience. To check for brain tumors and assess problems with brain anatomy, the magnetic resonance technique (MRI) was developed in the 1980s. This new technology method not only allows us to find anatomical problems, but also shows how the blood moves inside the brain in real time and in three dimensions. The brain has minuscule blood vessels called capillaries which carry the blood and oxygen to active brain regions. These blood vessels contract and expand in real time allowing blood to flow to different brain areas and it is possible to track with great precision which brain area is active at a given time. It is also a non-invasive technique, with no need for surgery or complex procedures. It was a revolution in studying the brain and thousands of studies followed.

At first researchers simply looked at which brain area was active when participants were shown specific types of stimuli such as images or sounds. Then, in 2000, researchers made a huge discovery. They realized that the brain is active even when people are not shown any stimulus. Researchers studying the brain's response to different tasks often use a period of rest or inactivity as a baseline but, during this period, they discovered that participants are not actually doing "nothing." Instead, they are engaging in various cognitive processes, such as mind wandering, self-reflection and planning. These findings suggest that the baseline period may not be a true resting state and this may have implications for the interpretation of brain activity during tasks. Although that would seem obvious, this was a major revelation for neuroscientists, and they coined the term the "default mode" of the brain in 2001. This is now called the default mode network because, when we are asked not to do anything, there is a network of brain areas active.

About 10,000 articles have been published on this subject. And yes, you have guessed right, this network of brain areas is also active when we have spontaneous thoughts and our mind wanders.

CHAPTER 4

HOW DO WE "TEST" MIND WANDERING?

It would be fantastic to have a mechanism to peer into people's minds and know what they think. As surprising as it sounds, we do have such a tool: it is called language – we can ask people what they are thinking! And this is what scientists do. They question people in their own environment or ask people to come to the laboratory and perform repetitive tasks when mind wandering is more likely to occur.

Before we delve further, it is important to describe the scientific method. Not all studies qualify as science. It is one thing to notice a pattern, and another to study it scientifically. Let's look at a spontaneous thought, like feeling bad about eating ice cream – which qualifies as mind wandering under the broad definition – and see if we can apply the scientific method to it. We could, for example, interview people at an ice cream booth, asking them the question: "Do you feel bad about eating ice cream?" You might ask the question informally to 10 people at random and 6 could answer positively. However, this is not science because it is not conducted and documented rigorously. Science is about reproducibility, so the experiment can be reproduced. The questions, for example, must be the same for everybody: "Before I talked to you, and after or during eating this ice cream you just bought, did the thought cross your mind that you regretted, felt guilty or bad about eating it?" The answer can be yes or no. If we collect 100 responses, we could estimate the proportion of people who have had a particular thought. Our results would likely depend on the geographical location and demographics of the population we are studying – the age, gender and education of those interviewed, so we must document this information.

Despite being simple, this would be valid science that could be published in a scientific journal. We might be able to know that, given a specific population and location, say 50% of the people spontaneously have negative thoughts associated with eating ice cream. Using the scientific method allows us to put some numbers on a given phenomenon, instead of speculating about it. A scientific report is also essential because it describes the methodology used for collecting data and preparing the results. Again, science is about reproducibility: based on the scientific report, other scientists should be able to conduct the same experiment and reproduce the same results.

What is going on inside our heads?

Cognitive scientists have developed fun ways, in laboratory settings and daily life, to study mind wandering and to find out what is really going on in our brains when we are not focused. Some studies probe our unconscious mental mechanisms by recording the answer to the question, "What are you thinking?" at random intervals during the day. Scientists may also ask people to press buttons to indicate the content of their thoughts in laboratory settings. These studies are in the domain of "experimental psychology."

Psychology studies the mind and behavior using surveys and naturalistic observation – the observation of people in their usual environment as opposed to the laboratory. Psychology researchers may ask patients to fill out questionnaires and make inferences based on their responses. For example, a clinical psychologist may give alcoholics and non-alcoholics a questionnaire indicating their experience during childhood, and find out that alcoholic individuals are more likely to have experienced childhood trauma

(Mirsal, 2004). By contrast, experimental psychology measures behavior instead of asking people to fill in questionnaires. For example, if you ingest alcohol, your ability to respond to an emergency while driving is impaired. Measuring the exact response delay – in a car simulator, for example – is part of experimental psychology.

Researchers also use neuroscience methods to study mind wandering – recording activity in the brain and other biological signals such as hearing rate, respiration, body temperature, skin conductance, etc. I am a Neuroscientist by training but I have also used survey methods (regular psychology) and recorded people's response times on a variety of tasks (experimental psychology). In this chapter, we will focus on experiments in experimental psychology to probe mind wandering.

Thought probes

A simple and standard mind-wandering experiment may consist of inviting you to a laboratory and asking you to play a rather dull car racing video game. Then, every one or two minutes, the game stops and the experimenter asks you what you were thinking about: "Were you on task? Did you think about something else? How long have you been distracted?" There may also be other questions that determine the content of your thoughts. Then the game resumes.

Note that, usually, a computer program, not a researcher, stops the game and asks these questions. These interruptions are called "thought probes." This type of experiment, and all that follow, may be conducted with or without other biophysical and brain activity recording. If no other sensor is used, researchers will study accompanying behavior in terms of speed and frequency of response. When brain and other physiological information is recorded, researchers can correlate the physiological activity with the observed behavior.

But are people truthful in their answers, you may ask? What if people lie about their mind-wandering episodes and ruin the results of the experiment? It is a risk that is important to address. People usually lie for a reason – for example, if they think they *must* stay concentrated during the experiment. Thus, the researcher needs to clarify for the participant that it is acceptable to mind wander, perhaps by explaining clearly the goal of the experiment: "Everybody gets distracted. If we can determine how much you are distracted during this simulated driving task, we might be able to improve driving safety so please try to be truthful in your responses."

Some questions might be more delicate than others. If you ask the question, "What were you thinking about?" some might not be willing to reveal personal thoughts. It is better to ask people to put their thoughts into categories such as:

1. Thinking about the task
2. Thinking about personal matters
3. Thinking about the future
4. Thinking about the past
5. I have forgotten what I was thinking about

This will also make it much simpler for researchers to analyze the data because they can count the number of responses in each category.

When we analyze data, assuming only a few percent of our participants were less than truthful, we might be able to determine which ones they are. To do so, we look for outliers. For example, we can look at the time it takes to answer questions when they pop up. If most people take one second, but a few take 1/10 of a second because they did not even read or think about the questions, we can remove these speedy answers. Even after accounting for outliers, some responses from people who lied or responded randomly will remain, and that's OK. This is considered "noise" in the data. This noise is likely to be similar when you compare conditions, for example, mind wandering while driving to mind wandering while washing the dishes. There is no reason why people should lie more in one case than in the other.

Self-caught mind wandering

So far, we have interrupted people when they are doing a task, asking them about their thoughts. Another way to determine if someone is mind wandering is to give them a task and ask them to report when they stop doing it or are distracted. As people must catch themselves mind wandering, these events are called "self-caught" mind-wandering episodes.

Self-caught mind wandering requires a relatively repetitive task like counting the breath. Ideally, the task must also be mental, so it isn't possible to mind wander and perform the task simultaneously. For example, if you are washing the dishes, you can usually do so and think about something else simultaneously. By contrast, if I ask you to count your breath backward, as in the exercise below, it will be difficult for you to do so and be able to think about something else at the same time. Demanding mental tasks are the type of task we would use to study self-caught mind wandering.

EXERCISE: SELF-TEST FOR MIND WANDERING

One task for self-caught mind wandering that I have used in my laboratory is to ask participants to close their eyes and count their breaths in reverse, starting from 10 (Braboszcz & Delorme, 2011). So starting at 10, then go to 9 as one breath is taken, then to 8 on the next breath and so on down to 1, and then start again at 10. The task is repetitive and relatively dull so our minds will spontaneously wander. I asked participants to press a button on their lap if they realized they had lost count of their breath because they were thinking about something else.

This type of experiment taps into spontaneous meta-awareness – the self-awareness of the thinking process – and I have used it to study what happens in the brain at the moment when a person becomes aware their mind is wandering. Let's try it.

1. Sit in a quiet place where you know you will not be disturbed, set a timer on your smartphone for 10 minutes and read the instructions for the exercise before starting.

2. You will try counting your breath backward from 10 down to 1, then starting again at 10. One breath is one breath in and one breath out.

3. When you have lost the count, start again at 10. Use your fingers to count the number of mind-wandering episodes. Your hands should be closed when you start, and you raise one of your fingers every time you lose count of your breath. If you have more than 10 mind-wandering episodes in 10 minutes, just restart your finger count at 1.

4. Thoughts will pop in your mind such as "I got it" and "I do not need to do the 10 minutes" but ignore these and stick to the 10 minutes.

5. If you are unsure about the count or hesitate, consider that a mind-wandering episode.

6. Now start the timer, close your eyes, and start counting your breaths.

This experiment is great to make you notice how hard it is to stay focused on something as simple as counting your breaths. It also shows how difficult it is to monitor your own mind. Did you feel any anxiety at the beginning? Did you notice the thought "I do not think my mind is wandering much" or "Am I doing the task correctly?" These thoughts are mind wandering as well, and not all mind wandering will cause you to lose track of your breath count.

Conducting experiments with self-caught mind wandering is now commonplace. A large number of mind-wandering experiments incorporate both self-caught and probe-caught mind wandering. In other words, there are probes at random intervals, but people can also report spontaneously when they are mind wandering.

The SART Task

The most popular of the psychophysics tasks to test mind wandering is the sustained attention to response task, or SART for short. This task is simple. You sit in front of a computer screen and are shown numbers every one to four seconds. Your job is to press the spacebar every time you see any number, except if it is number "3." So imagine seeing a blank screen, then a large number "5" appears for one second, and you press the spacebar; two seconds later, the number "6" appears. Again, you press the spacebar; three seconds later, the number "3" appears, and . . . you do not press the spacebar. Because the task is quite dull, you will likely get distracted. If you do this task for 20 minutes or so, we can determine when and how long your mind wanders by counting the number of 3s "missed" by pressing the space bar when they appear.

This task has several advantages: first, you have two measures of mind wandering in one task. You can count the number of 3s missed plus see how long it takes someone to respond. When people are distracted, even though they might be able to not react to the number 3, they usually respond more slowly to other numbers. If people respond slower, we know they are mind wandering, although not to the extent of failing to do the task. Secondly, this experiment is easy to set up and program on a computer and also easy to teach people to perform. Thirdly, because it has been used extensively, if you obtain some results on, say, alcoholic versus non-alcoholic individuals, you can compare them with other published studies. For example, you can compare it with a study using the same task with schizophrenia patients. When everybody uses the same protocol, it is easier to compare mind wandering in different conditions and mental states.

A downside is that we don't usually sit in front of a computer screen watching numbers appear one at a time so it is unclear if the results found with this task translate in the real world: if I use this task to show that young people's minds wander more than older ones, would that also apply to mind wandering while reading? Conventional wisdom says, "Yes, of course!" but scientists are doubtful. Note that in this particular example, it has been done, and the result on both tasks is similar (Jackson, 2012): younger individuals' minds wander more than older people's on both tasks.

The SART task has also been criticized because it does not involve higher cognition. The task is simple so some might be able to do it and think about something else simultaneously, which would invalidate some of the results obtained using this task. Finally, the task is so simple that it is difficult to probe the depth of mind wandering and whether participants were completely or slightly distracted.

If the SART task is limited, are there any other tasks that researchers can use to study mind wandering?

Other tasks to study mind wandering

The metronome response task

Another type of sustained attention task is the metronome response task (Anderson, 2021). As its name indicates, this requires people to follow a metronome beat by repetitively pressing a button. Researchers realized that the more people were off rhythm, the more likely they were to mind wander – the regularity of their rhythm indicated their state of mind. This makes it possible to assess the depth of the mind-wandering episode by calculating how offbeat people are. This task is often combined with thought probes (random questionnaires popping up on the computer screen) asking about the level of people's attention and the content of their thoughts. It is also possible to present these probes at the moment when we know the person is offbeat. We can only show so many thought probes (usually about one every 2 minutes), so it is essential for research purposes to present them at the right moments – when we know people are mind wandering – and the metronome task allows us to do precisely that.

Continuous Temporal Expectancy Task (CTET)

Another type of sustained attention task is called the Continuous Temporal Expectancy Task (CTET) (Irrmischer, 2018). In this task, you are asked to pay attention to images appearing on a screen for about one second and to press a button when you think the image lasts longer than the previous ones. While most images stay on screen for one second, about one in five images appear for a little longer, about 1.2 seconds. This task is difficult and requires attention. It also taps into a different type of sustained attention – a more involved type of concentration that is typically not possible when the mind wanders. Thinking about how long something lasts and being simultaneously distracted is difficult.

Finger tapping task

The finger tapping task also requires a different type of concentration. Here, people are asked to produce random rhythms with their two index fingers (Groot, 2022). Imagine two buttons in front of you, one for your left index finger and one for your right index finger. Now, generate a tapping sequence of one to three taps per second for each hand and make it as random as possible. Tap right – tap left – tap right twice and fast – gap – tap left three times – tap right once – being careful of not tapping the same sequence using both fingers. A computer program can easily determine in real time whether your finger tap sequence is random or repetitive. It can also determine if the tapping from both fingers is synchronous. Not surprisingly, the frequency at which people press the buttons and the randomness decreases when their minds wander.

* * *

All these types of sustained attention tasks tap into different attention mechanisms and use different types of cognitions: duration perception, movement synchronization, etc. Since scientists do not know which cognitive centers are most involved in mind wandering, some of these tasks might interfere more with mind wandering than others.

Carrying out these types of experiments was also problematic for epistemological reasons, i.e., what is politically correct in science. It reminds me of my first scientific experiment as a junior professor in 2005. At this time, I was a professor at Toulouse University, a university founded in 1229 and one of the oldest and most prestigious science universities in Europe. It was before mind-wandering and meditation research became legitimate scientific topics and I remember the tension in my Neuroscience department when I announced that I would study the brain activity of people's mind wandering during meditation – something I eventually did and even wrote a book about – the one you are reading now.

It was a taboo topic at this time, especially in a prestigious Neuroscience department. My colleagues were studying the brain response associated with visual and auditory stimulus processing, visual and auditory attention, spatial navigation, etc. However, researchers were not interested in meditation, and definitely not mind wandering, which most researchers had never heard of. My colleagues could not even grasp how someone would study such a topic. They were worried about my career and the department's reputation. Peer pressure was palpable. I did it anyway, and about a year later obtained funding to pursue my research which eventually led to the publication of a landmark article with my first Ph.D. student (Braboszcz & Delorme, 2011). The topic was now legitimate – the article had been published in a prestigious journal, and government agencies were willing to fund it. At the same time, other neuroscience laboratories in Europe and the US started to pick up on these topics. This emphasizes how, even in science where it is our job to push the boundaries of knowledge, it can be difficult to do something new.

CHAPTER 5

WHERE DO THOUGHTS COME FROM?

To study and simulate mind wandering, it is important to determine its origin. In the previous sections, we assumed that thoughts come from the brain "machine", and a large body of evidence supports that hypothesis. In 1950, neurosurgeon Wilder Penfield used brain electrical stimulation to generate random thoughts and hallucinations in awake individuals undergoing brain surgery. Because the brain has no pain receptors, surgeons can stimulate it mechanically or electrically after opening the skull, even when the patient is awake. To this day, this technique is still used by surgeons to check that they are not removing critical brain parts as they operate. For example, surgeons will stimulate a part of the patient's brain and patients will report the smell of flowers. They will stimulate another area, and patients will report thinking about a memory from their childhood. These experiments show a strong coupling between the brain functions and our thoughts.

Do thoughts come from the brain?

However, does this mean that all thoughts come from the brain? Some reports of telepathy are hard to ignore, and because we do not understand how a phenomenon could occur, we shouldn't just discard it. For example, one well-known story is about the "Jim Twins" – the twins were separated at birth but ended up marrying women with the same name, divorcing and remarrying another woman with the same name again, and naming their son the same, as featured in the *New York Times* (Chen, 1979).

There are dozens of similar stories and thousands of published reports, the most well-known being a study by Targ and Puthoff in the prestigious *Nature* journal (Targ & Puthoff, 1974), demonstrating that some talented individuals could read other people's thoughts. In this paper, they had a purported talented psychic draw the location of where one of the scientists was currently at, a location decided upon randomly. The psychic's performance produced way higher results than random selection. Over the series of experiments, there was a 1 in 10 million chance that someone with no ability would be able to guess the responses as the psychic did. Skeptics might argue that the experiments were flawed, however, the evidence points to the contrary. At the time, this was published in the most prestigious scientific journal – the same journal that publishes ground-breaking discoveries about the image of black holes or new elementary particles in physics. It was thus subjected to the highest level of scientific scrutiny. It does not make these phenomena real. They remain difficult to reproduce in laboratory conditions and involve unknown mechanisms. However, it would not be fair to discard them as pseudo-science.

The fact that thoughts could originate outside of the brain is currently ignored by mainstream science. However, it is essential in the context of mind wandering that scientific beliefs do not limit us. Human consciousness is poorly understood, as we will see in the next section. If thoughts can arise outside the brain, then the brain could act like an antenna capturing information and playing some of this on the "mind radio channel."

Suppose telepathy and non-local connection between brains is possible. It would not invalidate known results as neurons in the brain would still be active when some thoughts are active. Also, it would not change the dynamics of mind wandering, although it could revise its function. In that case, one of the functions of spontaneous thoughts could be to keep us connected to others, so we may, for example, feeling when our loved one is in danger, as has been reported for identical twins for example (Playfair, 1999).

The relationship between thoughts and consciousness

With the famous quote, "I think, therefore I am," Descartes was trying to find core principles that he believed were fundamental (Descartes, 1641). However, any meditator would know this sentence is not entirely accurate. I interpret this sentence as, "I think, therefore I know I am conscious, and therefore I am." It could also be construed as, "I perceive the world; therefore, I am." However, "I think, therefore I am," is misleading because it suggests that if I do not think, then I do not exist, which is not true. When we experience something, we are often not thinking. Take the experience of watching a beautiful sunset. For some time, you might be in awe and you are not thinking of anything. Your mind is not blank, but it does not contain thoughts, just the experience of what is happening at that moment.

Consciousness is the entity who knows, and the content of our consciousness is our thoughts, feelings and perceptions. Consciousness may be seen as a container of thoughts, although the distinction between container and contained is not always clear, as our conscious mind often believes it is the very thing it contains. Upon thinking, "I am so smart," we do not usually experience it as a thought passing through our mind but instead embed it and believe it is what we are.

It is the same with perceptions – when our mind wanders, we sometimes experience the distinction between container and contained because we notice that an idea has hijacked our mind. What is consciousness, then, if it is not thinking? This is important for our understanding of mind wandering – when our mind wanders, are we wandering ourselves, or is there something separate that contains the wandering thoughts?

Consciousness is an illusion

Based on what we know, the current scientific stance is that there is no container to mind wandering: mind wandering contains itself and consciousness is a byproduct of the brain and, consequently, an illusion. In this model, as a person you have no free will. You believe you do, but these beliefs are just thoughts popping up in your brain. Neuroscientists like to study split brain patients, where the brain has been severed in the middle to prevent some forms of extreme epilepsy. If you show an image on one side of the visual field, then only half of the brain sees it. If you show an image to the right brain with some instructions (laugh; walk), then the right brain – which can read – understands and the person starts acting out the instructions. However, the left brain has no idea why. Only the left part of the brain controls language centers, so if you ask the person why they laugh or started walking, they

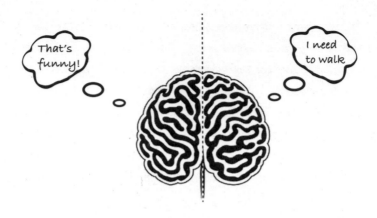

will make up a story: "I laughed because I thought the experiment was funny," or "I walked because I wanted a soda." This is one reason some neuroscientists think consciousness is just byproduct phenomenon: our brain takes a decision and then we become aware of a story we have made up. In this model, your brain is a biochemical supercomputer capable of taking care of your body and making decisions – being conscious makes no difference because consciousness is unnecessary.

We are circling back to the hard problem of consciousness I mentioned earlier. We believe we exist, but how can dead matter give rise to consciousness? Scientists do not know. There are hand-waving explanations, but none are satisfactory. If we knew, we would be able to design and maybe even build a conscious robot or computer. Despite some of the headlines in mass media about the marvels of new artificial intelligence systems, scientists are as far as they have ever been from this lofty goal.

Consciousness as a field

There are alternative hypotheses where mind wandering happens in consciousness. I like to think of consciousness as a forcefield and the brain as a device to confine that field. The consciousness field could be a property of nature in the same way that electrical fields are. Because it is a field with infinite reach, electricity is everywhere. Its amplitude varies with distance, of course. If I move a magnet, I can measure its induced electrical field. With some precise instrument that does not exist yet, someone may also be able to measure that same field several miles away. The electric field extends to the other side of the galaxy, and in theory, to infinity. Yet, a couple of hundred years ago, our ancestors had no idea it existed. Could consciousness obey similar rules? Could the field of consciousness permeate space as electricity does and only be concentrated in devices called brains, like electricity is contained in batteries? The advantage of theories such as this is that they solve the hard problem of consciousness. Consciousness does not have to arise from dead matter because it is its own thing. What are the consequences for mind wandering with this theory? They are mind boggling. The main property of consciousness is to "know." Consciousness knows perceptions, knows thoughts, knows itself. Consciousness has relational properties. Now, with the understanding that consciousness is the container of our thoughts, we can decide to see thoughts as they are: transient impermanent activity happening within us.

* * *

We have seen already that thoughts either influence or originate in the brain. fMRI scanners can "see" thoughts in real time as someone watches a movie, and other machines can determine from the brain's electrical activity what sentence someone is listening to. Future research will determine if we are able to design devices that influence thoughts as they appear in consciousness – without changing who we are. We may make new types of thoughts appear, counteracting existing negative mind wandering. We may be able to cure people with pathological mind wandering or influence the course of a disease.

PART 2

MIND WANDERING AND YOU

CHAPTER 6

WHAT ARE WE MIND WANDERING ABOUT?

What topic do our minds wander on? What does this tell us about our mind-wandering habits? Do our thoughts influence future mind-wandering content? How can these thoughts be better controlled?

Remember the different types of mind wandering discussed earlier, such as spontaneous thoughts, intentional thoughts, stimulus-dependent and task-dependent thoughts, and obsessive thoughts? The content of stimulus-dependent and task-related thought clearly depends on the environment we are in or the activity we are doing, but what about the origins of the other types of mind-wandering thoughts?

When our minds wander, it seems that we often think about ourselves and about others, including partners and friends – imagining ourselves replaying events, planning future conversations, or fantasizing about hypothetical exchanges. How do scientists know this? They use the tests we have described previously, interrupting participants with questionnaires to probe their thoughts, or letting them work on a repetitive task and spontaneously reporting their self-caught mind-wandering episodes. For example, a questionnaire may pop up asking participants: "When/If you were mind wandering, what were you thinking about? Check all that apply:"

- ✓ Yourself
- ✓ Others
- ✓ Nobody
- ✓ The past
- ✓ The future
- ✓ It was positive
- ✓ It was negative

This type of study is done by sending text messages to people throughout the day and asking them if they are concentrating on the task they are doing and, if not, what they are mind wandering about.

Mind-wandering content

Studies show that we are thinking about ourselves in relation to others in 45% of mind-wandering episodes. For example, we may spontaneously think of what we will tell our partner when we get home. In another 40%

of the cases, we are the only agent (Smallwood, 2015). This includes thoughts about plans for our day, our week or our life. This is when we also make resolutions such as "I need to lose weight." In the remaining 15% of mind wandering, we think about nobody in particular. For example, we might be thinking about philosophical or practical issues that do not involve others.

What about past versus future? The same study discovered that about 50% of our mind wandering is about the future, 30% about the present, and only 10% about the past (Smallwood, 2015) – and 10% has no sense of timing at all. Note that this is not the case for depressed people. These individuals have a much larger number of thoughts about the past, and the proportion of negative mind wandering also increases for them. Also, interestingly, mind wandering about the past is generally more negative than mind wandering about the future. In the questionnaire above, when participants checked "The future" box they were more likely to also check the box "It was positive" than when they checked "The past" box.

How content evolves

In another study from the University of Miami (Zanesco, 2020), researchers asked more than 500 participants to indicate the content of their thoughts when their minds wandered. As in most mind-wandering experiments, they asked participants to do a repetitive task and interrupted them at regular intervals to ask them what they were thinking about. In this experiment, those taking part had several choices and they could only pick one. The options were:

1. Being focused and on task
2. Having thoughts about the task they were doing
3. Other thoughts

For other thoughts, participants were offered additional choices although it is not relevant here.

The figure below shows how these choices vary over time (Zanesco, 2020). Over the course of 13 minutes, people are interrupted multiple times and the average content of their thoughts is shown. Every vertical line shows the contents of the participant's thoughts in that time period. At the extreme left, which corresponds to the beginning of the experiment, we see that most participants are either on task or have task-related thoughts. The dark colors in the lower portion of the figure correspond to these two conditions, representing more than 95% of all responses. However, at the extreme right, which corresponds to the end of the experiment, these two responses only account for about 40% of all answers. Then the most significant category was non-task related thoughts about people's current physical, cognitive or emotional state. So this illustrates how people's thoughts evolved throughout the experiment.

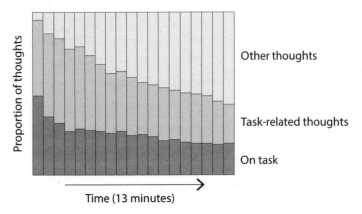

Time (13 minutes)

Frequency of mind wandering

Since mind wandering can be involuntary, assessing how long mind-wandering episodes last isn't easy. It is important to know because our goal might be to reduce the duration of mind-wandering episodes. Most reports on this topic are based on how long people thought their minds wandered, not actual time measurements.

In one study, to assess if mind wandering might change throughout the day, researchers sent text messages to study participants five times a day for 14 days. They discovered that the mind wanders more following meals (Smith et al., 2018). After eating a meal, the stomach diverts some of the body's energy from the brain, which is why we sometimes feel sleepy. Mind wandering occurring more after meals is consistent with the idea that it is a state of microsleep that may help us rest after taking a meal.

As with dreams, we may not always remember the content of our mind wandering. It could be that our mind simply goes blank, a state some researchers call "mind blanking." However, it could also be that we might fail to remember the mind-wandering content. One of my students had been trained to conduct a thorough interview to probe people's thoughts and we decided to trial it on mind wandering, and I was her first subject. I did a repetitive task and waited until I realized my mind had drifted. After that, she interviewed me for 20 minutes on this single mind-wandering event, asking me to recall my detailed impressions. It was like trying to describe a dream. I did remember some of it, but not all. Also, as I was describing it to her, I started remembering more.

Is mind wandering gradual or "all or none"?

To answer this question, we must determine whether mind wandering can be partial or if it only has two states – mind wandering and not mind wandering. In other words, can your mind wander at 50% with 50% of your mind still engaged in an activity, or can it only wander at 100%?

Researchers have attempted to study this question. They typically run these studies by interrupting research participants and asking them at what level they were either on task or thinking about something else, and to grade their experience from 1 (completely on task) to 5 (completely off task) (Zanesco, 2020).

People used all values (1, 2, 3, 4, 5), indicating a gradient of mind wandering. However, the data analysis revealed something remarkable about the completely on-task and off-task mental states – the "1" and the "5." In these mental states, participants were more likely to remain in that state than to transition to an intermediate level of being on task. These states were more "sticky" than others.

A large statistical analysis of many mind-wandering episodes points toward three mental states: focused on the task, mind wandering about the task or everyday event, and deep mind wandering or daydreaming. Why is this important? It is important for safety because it means that sometimes we can mind wander but still do the task. However, we might not be able to do it to our full potential. Would you want an airline controller or pilot trying to do their job while being distracted, thinking about something else at the same time? This is why there is a lot of mind-wandering research in aviation safety.

Does content influence future mind wandering?

It's also useful to know how thoughts appear in succession. Do previous mind-wandering thoughts influence the content of subsequent ones? For example, if your mind wanders about a future vacation, are you more likely to think about it again later? What is the time scale? Does a mind-wandering episode now influence others for minutes, hours or days?

Interestingly, controlled studies show that people often report thinking about the same thing for two interruptions that follow each other (Zanesco, 2020). So, for example, if your mind wanders about a personal worry once, it is likely that it will do so again shortly after. Even if one mind-wandering thought can only influence the next, though, can this still have a ripple effect? Does one mind-wandering thought influences the following one, which affects the following one, etc.?

Some researchers believe that mind wandering involves long-term memory and can influence many future episodes. The analysis of many mind-wandering events can help determine which one is more likely. Let's imagine how your mind works – it has your list of mental worries, everyday concerns and other topics of interest, and it randomly jumps to one of these when it wanders off. What does this mean for you and me? It means that you need not worry about your past thoughts. Having a negative thought about the present moment, in the present moment, does not dramatically increase the likelihood of having negative thoughts in the future. The likelihood of a thought popping up in our mind is relatively independent of what happened before.

The following graphics illustrate that it is possible to create a model that only remembers the previous mind-wandering episode and reproduces a similar sequence compared to people's responses (Zanesco, 2020). There are three mental states: on task, partially on task/partially mind wandering, and mind wandering. Every second, the mental state is re-evaluated so there is a chance of the changing of mental state.

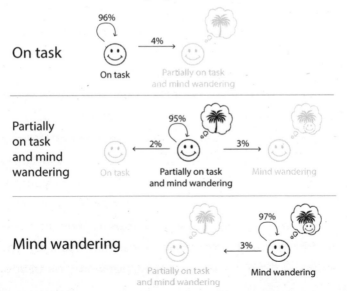

When we are in a given state, every second brings a probability of staying in the same state (circular arrow) or jumping to another state. The likelihood of transitioning from one mental state to another is calculated based on a large number of responses from people, and this model will produce mind-wandering episodes comparable to a real person. So, if we are in the on-task mental state, then the next second, we have a 96% chance to be still on task and a 4% chance to transition to the "partially on task and mind-wandering" mental state. In this state, if asked, we would respond that we have task-related thoughts or think of everyday things (lunch, what's next, etc.). Also, you are continuing to perform the task in this state. Even though 4% of transitioning is not much, because it happens every second, after 15 seconds, if we start "on task," we have about a 50% chance of transitioning to the "partially on task and mind-wandering" state.

Once in "partially on task and mind wandering," the probability of staying in that state is 95% and we might remain there for a while. However, after 30 seconds, there is a 60% chance we will transition to the "mind-wandering" state and a 40% chance to return to the "on-task" state. The "mind-wandering" mental state is very sticky. It has the lowest probability of transitioning to another state (3% only), so you're likely to stay in this state longer than others. At some point, though, you transition back to the "partially on task and mind-wandering" state. You will notice that there is no direct transition from "on task" to full "mind wandering," because it only very rarely happened in the data collected on the 500 participants. The data shows that people rarely jump from a state of high focus to a state of mind wandering, but instead go through the intermediate state.

People also differentiated the two types of mind-wandering mental states. Full mind-wandering thoughts were more unpleasant, out of control, strange, racing and suspicious than partial mind wandering pertaining to the task or everyday events. So although partial mind wandering is not necessarily positive, it is not seen as negatively as full mind wandering.

* * *

What does knowing about the content of mind wandering teach us and why does it matter that we understand transitions between mental states and mind wandering? It is important because, when we try to curb our mind wandering, we can take into account common subjects on which we mind wander or understand that deep mind wandering seems to be preceded by partial mind wandering about the task or everyday events. We can devise strategies where we observe our minds and assess if we have any thoughts about the task at hand and we may become aware of patterns of thought sequence (such as partial mind wandering then deep mind wandering) simply because we know research has shown them to be present in other people. This way, we become aware of the patterns of our own mind wandering ahead of looking to make changes in our lives.

CHAPTER 7

THE NEUROSCIENCE OF MIND WANDERING

We have mentioned many mind-wandering experiments based on people's behavior, but what about the brain? What happens there when our mind wanders? There are a large number of experiments involving brain imaging that teach us about mind wandering. These show us that there are different types of specialized brain regions dedicated to different parts of human cognition, and spontaneous thoughts interact with each of them differently, as we can observe in real time using brain-imaging techniques.

When we understand which cognitive systems are involved in mind wandering, we not only understand how our mind works intellectually, but may also be able to detect patterns in our mental activity that we did not know about. Becoming aware of these automatisms is the first step to being able to change them.

This section contains a lot of names of parts of the brain which can be confusing. The terms are not important; what is essential is the function of these brain areas. If I were to explain how an engine works, we would need to talk about the carburetor and piston, and so it is with the brain. We need to use the correct terminology yet it is the mechanisms that matter.

Inside the brain

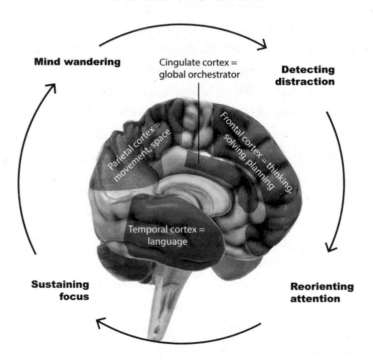

Mind wandering

Cingulate cortex = global orchestrator

Detecting distraction

Parietal cortex = movement, space

Frontal cortex = thinking, solving, planning

Temporal cortex = language

Sustaining focus

Reorienting attention

A few guiding principles can help you better understand the terminology of the brain areas we will discuss, as shown in the figure above. The human brain is divided in different parts: the cortex is the outer layer of the brain that is responsible for many of the brain's cognitive functions, such as perception, thinking and voluntary movement. The **frontal cortex**, which is part of the front of the brain, contains many different subparts, and most involve remembering, problem-solving and planning. We will also talk about the **parietal cortex**. This cortex is involved in moment-to-moment movement planning and is usually active if you are getting ready to redirect your attention or perform a movement. Then we have the **temporal cortex** located, as its name indicates, behind the temples. It is the part of the brain involved in language.

Finally, we will also talk about a part of the cortex folded inside the brain's center called the **cingulate cortex**. You may notice that the brain depicted above is not standard. We have cut part of it away so we could see inside. Cingulus means "belt" in Latin, and scientists probably chose the name because this cortex surrounds the brain's center. The cingulate cortex can be seen as the global orchestrator in the brain – it is central so has an ideal position to communicate with and keep all other brain areas in check.

Sustaining focus

The brain area most involved in maintaining attention is the dorsolateral prefrontal cortex, which is part of the frontal cortex. This cortex is active when concentrating on one task. Many other areas are involved when we focus, but this is the most engaged.

This type of cortex is also heavily occupied with short-term memory. If I ask you to memorize a phone number for a few minutes, this involves short-term memory, which is also needed for problem-solving since you must remember the problem and relevant factors in order to solve it. For example, if you are thinking about whether you have enough money in your bank account to make it to the end of the month, you will try to remember what you still have to buy and perform mental calculations using this cortex. This cortex is similarly active when you meditate and are not solving any task, because you still need to focus your attention and remember your meditation instructions.

Why is it useful to know that this brain area is active when we sustain focus? Well, it is possible to use neurofeedback to know in real time that this brain area is active, then work to sustain the activity in this brain area longer and train the brain to be focused for longer periods of time, as shown in the image on page 94.

The character is trying to control the color of a square with his mind by controlling the activity in his frontal cortex. The color gets darker when they concentrate hard. We will come back to this experiment later on when we talk about brainwaves and neurofeedback.

Starting to mind wander

From the "Sustained focused" state, we may drift to the "Mind wandering" state as shown in the brain diagram above. When our mind starts wandering, this involves a second type of attention that deals with "conflict monitoring" – the attentional system that triages multiple objects competing for conscious access. The "brain attentional system" is a group of brain parts that work together to help you focus your attention on what you need to. For example, we may hear our favorite song while reading. Should we listen to the music or continue reading? These activities compete for our attention, and the conflict-monitory attentional system will help us decide which one to attend to. This type of attention is present during mind wandering as, when our minds wander, at least two objects are competing for our attention: the task we are supposed to do and our spontaneous thoughts.

The Default mode network

Researchers have observed part of the default mode network we mentioned in Chapter 3 being activated during mind wandering. The default mode network is involved in all mind-wandering episodes, but its activity is maximal when subjects report being unaware of their mind wandering. Remember that the default mode network is one of the active brain areas when we are at rest, not interacting with the outside world. Many brain areas are part of the default mode network: the medial prefrontal cortex gets active when we make decisions. Another part of the default mode network is the cingulate cortex, folded inside our brain, which is linked to self-awareness and decides where we allocate our attention. Finally, a part of the temporal cortex is involved in language comprehension. All of these brain regions talk to each other when our minds wander. Why? We are usually thinking in words (temporal cortex), planning (frontal cortex for planning), as well as being aware or unaware of our thoughts (cingulate cortex). The default mode network is much more than mind wandering, but for the sake of simplicity, we will equate default mode network and mind wandering in this chapter.

Also, let's not forget that memories are involved during mind wandering. Where are memories stored?

Everywhere. Memories are embedded in the connections of neurons within and between cortical regions. When our minds wander, it is usually about the past or the future. The default mode network is involved in recalling memories and relating them to the current situation so is active when we imagine being in a different place or time (Andrews-Hanna, 2010). This is called "mental time travel," and it often happens when our mind wanders.

It will likely be possible in the future to directly influence the activity in each of the brain areas described above. The technology exists and is called transmagnetic cranial stimulation (TMS). It is not precise enough to target specific brain regions yet, but it will be at some point. For example, imagine we are able to build a mind-wandering detector. This detector could send a signal to a TMS machine that would stimulate the temporal language area and disrupt the silent narration associated with mind wandering, so helping us refocus on the task or even avoid losing focus.

Detecting distraction

Once we start mind wandering, how do we become aware of it? As we have seen above, even when mind wandering, our spontaneous thoughts compete for our attention with the task we are doing. At some point, we realize we are mind wandering. This realization happens in the **salience brain network**. A brain network is a group of brain regions that are functionally connected and work together to carry out a specific cognitive or behavioral function. The regions in a network communicate with each other through neural pathways, and the activity in one region can influence the activity in other regions within the network. The salience brain network aims to detect novel – potentially salient – stimuli and then signal to the rest of the brain that we might want to redirect our attention, e.g., as we hear a bird, we might want to make eye contact with it. The salience network supports us in *noticing* the bird – it also detects that our mind is wandering. It is not shifting attention (yet) – that is handled by another network, the executive brain network, which we will deal with in the next section.

Meditation and the salience network

The regular practice of meditation induces long-term modifications in the brain. Long-term meditators can decrease the occurrence of mind wandering, which is supported by increased activity in the salience network (Brewer et al., 2011). Scientific studies show that meditation practice enhances connectivity between brain areas within the salience network, and between this network and other brain networks, making it easier for this network to detect distraction (Tomasino, 2014).

The increased connectivity is also observed when meditators are shown negative emotional images, indicating that emotion and the capacity to avoid distraction are related. Sustained activity in the salience network suggests hypersensitivity to disturbing images and mind-wandering episodes. In other words, expert meditators' minds might not wander as much because they can detect spontaneous thoughts as they form (Taylor et al., 2013).

So does it mean we must increase activity in our salience network to become more aware? Could we again use neurofeedback to train ourselves to increase activity in this brain region, and be able to detect mind wandering and become a super aware being? Not so fast. Mind wandering also occurs in depression and anxiety. A study found that the salience network showed increased functional connectivity during sustained negative thought compared to neutral or positive thought conditions. The study also found that this increased connectivity in this area was related to anxiety. As I will argue later, it might be because anxious people are caught in a cycle of negativity. For them, the salience network might detect that you have a positive thought and signal it to the rest of your brain to be replaced with a negative one. Increasing activity in this brain area with neurofeedback might not achieve the desired effect unless we are able to detect the type of mind wandering (positive or negative) the person is experiencing.

Regaining focus

Finally, once our mind has detected a mind-wandering episode, our brain needs to redirect our attention back to our primary activity. The **central executive network** handles this. As its name indicates, this part of the brain controls the actual execution of the reorientation. To simplify: the frontal cortex instructs the parietal cortex to shift attention back to the task. We close the loop and return to a sustained concentration task until we slip again into mind wandering.

As we know, there are two key methods scientists use to detect mind-wandering episodes – interrupting people at random intervals or waiting for them to report their attention had drifted. Doing this while the brain is being monitored allows scientists to see activity supporting what people reported. Using both methods to detect mind-wandering episodes, researchers observed increased activation in the frontal cortex, which was involved in detecting mind wandering and redirecting our attention.

Awareness of mind wandering

When do we become aware we are mind wandering and is it important? Based on current research, we are more likely to become aware our mind has wandered when we are ready to redirect our attention (executive network) than when we detect a conflict (salience network). Why? Research has found that people who have stronger connections within the brain areas of the executive network are more aware of their dreams. The executive network is responsible for cognitive control, decision-making and self-awareness, among other things. Therefore, the greater the internal connections within this network, the more likely people are to be aware of their dreams and by extension of their mind-wandering episodes.

It is possible to influence the activity of neurons in these regions using neurofeedback or TMS to help people become aware of their mind wandering. Which step in the cycle of mind wandering is easier to target? Based on our current knowledge, it is easier to influence the brain when focusing or when the mind wanders – the first two steps. Brain activity during the on-task and the mind-wandering periods is more sustained than in the last two steps (detection and redirection) – they last longer – so they are easier to detect using brain neuroimaging techniques.

What about brainwaves?

What are the brain-imaging methods that allow real-time peeking into the brain? In the previous section, the analysis of the active brain areas involved during mind wandering was based on functional magnetic resonance imaging (fMRI). Clinicians use fMRIs to detect brain tumors or other brain anatomical issues and even measure blood flow. The disadvantage of this method is that it is slow, not because of the technology but because blood flows slowly.

Another method is electroencephalography (EEG) – placing electrodes on the scalp. All neurons in the brain generate electrical activity and this technique detect changes in the brain's electrical activity. We can simultaneously use up to 512 electrodes and record the changes on these electrodes in real time. Compared to fMRI, the advantage of this method is that we are not limited in time precision, and it can collect brain activity thousands of times per second.

Brainwaves

You have probably heard about brainwaves – delta, theta, alpha, beta, and gamma are different types of brainwaves produced by the brain neurons' activity that correspond to brain oscillations at different frequencies. Delta waves (1 to 4 oscillations per second or Hz) are typically observed during deep sleep, and may also be present during states of unconsciousness. Theta waves (4 to 8 Hz) are a little faster and are linked to relaxation, creativity and memory. Alpha waves (8 to 13 Hz) are even faster and indicate a relaxed yet alert state of mind. (Alpha is the first letter in the Greek alphabet and was the first brainwave discovered.) Beta waves (13 to 30 Hz) are fast and occur during periods of mental activity and focus. Finally, gamma waves are the fastest (30 to 100 Hz) and are related to high-level thinking and attention.

What can brainwaves teach us?

Now that we have introduced the EEG technique, let's see what we can learn about mind wandering through this technique. Remember the experiment I designed asking people to silently count their breaths backward from ten to one and to press a button when they realized they had stopped counting their breaths because their minds had wandered? In this activity, the time before the participants pressed the button was the time during which their minds were wandering (Braboszcz & Delorme, 2011). The time after they pressed the button corresponds to a returned concentration on breathing. We observed that mind wandering was associated with an increase in slow waves (delta and theta) and a decrease in faster frequency waves (alpha and beta). The same increase in slow waves (delta and theta) plus a decrease in faster frequency waves is observed during sleep. These results led us to conclude that mind wandering is a state of low alertness, maybe similar to sleep. The brain is in idle mode, resting.

While participants were doing this task, we played beeps on loudspeakers every second that we instructed participants to ignore as shown in the figure below.

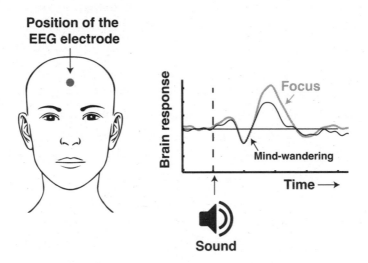

We noticed that the brain response to the beeps was smaller when their minds wandered. The brain had become less responsive to sounds. Again, this is similar to what happens during sleep. These results shed light on the potential functions of mind wandering as a state of rest for the brain, not quite like sleep but close.

Meditation, mind wandering and EEG

What about changes in EEG due to meditation practice? Meditation practices aim to maintain attention on a specific object and train the brain to detect mind wandering. They decrease mind wandering, as we have seen, so how meditation changes brainwaves could teach us something about mind wandering.

This is a complex subject. First, there are many different meditation traditions, and their effect on the EEG signal differs. A mantra-based meditation, where people repeat a word or sound to aid concentration, might not have the same effect on the brain as a breath-focused meditation or meditation on body sensations. I observed this when I studied the brainwaves of practitioners in four meditation traditions (Braboszcz et al., 2017). I conducted these experiments because, at that time, there were a lot of headlines about meditation's effects on the brain, but people meant different things by meditation. We wanted to verify that different types of meditation had similar effects on the brain.

I observed that the Vipassana meditation practices dramatically increased alpha brainwave amplitude, while other meditation traditions did not. Vipassana meditation is a type of meditation in which people do mental body scans, concentrating on each part of their body in turn. This type of spatial attention is likely linked to the increase in alpha brainwaves. Some meditation practices, such as mantra practices which consist of repeating a word or sentence in one's mind, also affect the brain in slower theta brainwaves in specific brain regions.

These results are consistent with meditation changing both slow and fast brainwaves, although the mechanism might differ across meditation traditions. As seen in the previous section, mind wandering increases slow sleep-like brainwaves and decreases faster ones, and meditation might be a way to alter that balance. Theta is associated with memory and cross-talking between brain areas. The alpha brainwave, and its absence, are correlated with attention. For example, if I ask you to track an object, the part of the brain that handles vision for that specific object will have decreased alpha brainwaves (Zani et al., 2020).

One change common in those who practice meditation traditions is the increase of fast gamma brainwaves. Meditators have globally more gamma brainwaves than non-meditators in all parts of their brain. Researchers also observed that gamma brainwaves increased when meditators were meditating compared to resting (Cahn et al., 2010). As soon as meditators started meditating, the gamma brainwaves increased, and over time this gamma increase seems to stay even after they stop meditating, which is why meditators have these higher levels of gamma brainwaves than non-meditators. Gamma is the brainwave of local processing within the brain – because it is so fast, neurons active in these frequencies only influence neighboring neurons. A gamma brainwave increase indicates increased local brain communication. Although researchers have not yet confirmed the link between gamma and mind wandering, increased communication within brain areas could be linked to a higher awareness of mind-wandering episodes.

The gamma results for meditators are also supported by brain anatomy studies showing a global increase in the thickness of cortex in meditators, a robust result reproduced by at least a dozen laboratories (Lazar, et al. 2005). This indicates improved communication within brain areas – increased thickness of the cortex represents more local connections in key brain areas involved in mind wandering. These additional connections could support greater and swifter awareness of mind-wandering episodes in the salience network. This hypothesis is consistent with the fMRI results showing increased activity in the salience network for meditators.

Brain-imaging studies have revealed that different specialized brain regions are involved in mind wandering and that spontaneous thoughts interact with each of them differently. We also showed that understanding which cognitive systems are involved in mind wandering can help us become aware of our automatic thought patterns as the first step to potentially changing them. There are also different brain-imaging methods (such as fMRI and EEG) that allow us to study the brain in real-time and gain insights into how it functions. EEG studies have shown that mind wandering is associated with an increase in slow waves (delta and theta) and a decrease in faster frequency waves (alpha and beta), similar to what is happening during sleep. This suggests mind wandering is a state of low alertness and rest for the brain.

CHAPTER 8

MIND WANDERING, HAPPINESS AND PERSONALITY

The previous chapter reviewed how the brain works and what it shows us about mind wandering. Let us now dive into some actual experiments and see what they teach us about ourselves. By becoming more aware of our thought patterns and behaviors, we can learn to identify and manage mind wandering, leading to increased self-awareness and well-being. We will see how our relationship with mind wandering links to our personality, emotional health, happiness levels, age and memory. Furthermore, I will describe how mind wandering has been linked to negative emotions such as anxiety and depression.

Studying mind wandering means we can develop strategies to improve our mental health as presented in the final section of the book. Mind wandering can also interfere with our productivity, but by learning about it, we can minimize its effects.

Your mental state also influences the content and frequency of the mind-wandering episode. An anxious, sad and stressed mental state leads to increased mind wandering. The strongest predictors of increased mind wandering are anxiety and boredom. This makes sense, as the more bored you are, the more your mind is occupied with spontaneous thoughts.

Children and adults with Attention Deficit Disorder (ADD) also have more frequent mind-wandering episodes, meaning that the probability of drifting from "on task" to other states is higher. By contrast, engaging in preferred activity tends to decrease mind wandering, reducing the probability of drifting from the "on task" state. This is consistent with the landmark study showing that mind wandering tends to decrease perceived happiness (Killingsworth & Gilbert, 2010).

Mind wandering make us unhappy

One of the seminal articles in the field of mind-wandering research, which we will call the "Harvard happiness study" because the researchers were from Harvard University, was published in 2010 and titled "A wandering mind is an unhappy mind," (Killingsworth &

Gilbert, 2010). For the experiment, Harvard scientists sent text messages to participants, asking them throughout the day what they were doing and if they were concentrating on that activity or thinking about something else. The text messages contained a link, and participants had to click on the link and answer a few questions such as:

1. What are you doing right now?
2. Are you thinking about something other than what you're currently doing?
3. If so, what are you thinking about?
4. How happy or unhappy are you at the moment?

Participants received a couple of text messages each day over a month. At the end of the month, they received a summary analysis of their responses. Researchers collected data on 2,000 participants and published the most recognized result on mind wandering.

Results of the Harvard happiness study

Before I show you the results, let me share my connection with this experiment. Shortly after the article was published, I signed up to be a participant. Even though the first phase of the data collection had ended, the researchers continued to acquire data for subsequent studies. On joining, I received one to three text messages a day asking me if I was focused on what I was doing. I could be doing anything – biking to work or having dinner with friends and I would receive a text message. I would stop, look at the text message, click on the link, take maybe 15 to 30 seconds to answer the questions on my smartphone, and then resume my activity.

Unpleasant mind wandering

Neutral mind wandering

Pleasant mind wandering

Not mind wandering

Subjective happiness rating

The results of this study were unequivocal. When our mind wanders, we are unhappy, as visible in the figure above. You can see people's subjective rating of happiness when they realize they were mind wandering while doing specific activities. The researchers captured people's minds wandering during many activities, including work, play, household chores, sports, watching TV, taking care of kids, etc. If you look closely at the diagram, you will see that people even took the time to respond during or after sex.

Irrespective of the activity, people reported lower happiness levels when their minds were wandering. It was true even when people were thinking about pleasant experiences (for example, the prospect of a vacation).

The researchers also found that what people thought was a better predictor of their happiness than what they were doing. So let's say you are doing something that is considered pleasurable – for example, eating ice cream. At the same time, you are thinking about problems at work, which makes you unhappy. I can predict that you are not very happy even though you are in the middle of a pleasurable activity. So, to assess how happy someone else is, we will have better luck if they tell us what they are thinking of rather than what they are doing.

This one-page journal article, published in the prestigious journal *Science* (Killingsworth, 2010), shaped the field and fuelled thousands more pages on the subject. It was groundbreaking in many ways. First, other scientists were able reproduce the results, a critical step in science. As surprising as it sounds, many studies, sometimes even high-profile ones, cannot be replicated – often due to a flawed experimental design or statistical analysis. This study was not one of these, and follow-up studies quickly followed.

These additional studies validated the original research and brought in new results that enhanced our knowledge of how mind wandering affects us. For example, researchers showed that mind wandering decreases and subjective happiness increases with age (Jackson & Balota, 2012). Older individuals reported less mind wandering and more pleasant thoughts when their minds wandered than younger individuals. This last study was performed with the sustained attention (SART) task mentioned in Chapter 4, so researchers were able to replicate results with a different methodology than the Harvard study.

Personality and mind wandering

Another important and revelatory topic in mind-wandering research is the relationship between mind wandering and personality. Studies have shown that about 50% of a personality can be attributed to genetics (Jang et al., 1996). How do we know? Researchers performed personality tests with twins growing up in different environments. Any common personality trait they share therefore can be attributed to their genes – nature rather than nurture. This is important if a propensity to mind wander is likely also determined partially by your genes – no twin study on this topic has been conducted yet but there are others as we will see below.

The "big five" traits

Psychologists usually consider five major personality traits: agreeableness, conscientiousness, extraversion, openness and neuroticism. The scale used to assess these personality traits is called the "big five" personality questionnaire – and there are shorter and longer versions. Many other personality assessment methods exist, but this is the most popular method to evaluate how personality is related to mind wandering (Muller et al., 2021).

Before we look at the connection between personality and mind wandering, we will look at the questions for the short "big five" questionnaire, and even do a simplified version of it so you can see where you fit in that picture. First, we will describe each of the personality traits, and then look at how these correlate with mind wandering. If we know that mind wandering increases with a given personality trait, that can help us devise specific strategies to curb mind wandering for this type of person.

EXERCISE: ARE YOU AGREEABLE?

The first personality trait of the big five is **agreeableness**. Agreeableness is about trust, altruism, kindness and affection; it is also about social harmony. Researchers have found that women score higher on the agreeableness scale than men.

To test yourself, consider the two statements:

1. "I consider myself as someone who is generally trusting."
2. "I consider myself as someone who finds fault in others."

For each statement, give yourself one point where you disagree strongly, two points if you disagree a little, three points where you are neutral, four points if you agree, and five points when you agree strongly. The sum of the score for the first question minus the score for the second question is your "agreeableness" score. The higher the score, the higher you are on the agreeableness scale. So to obtain a high score, you must trust other people easily (high score, such as five on the first question) and not find faults in others (low score, such as one on the second question). Then the difference between the scores is highest (four in that case). A score of zero would indicate you are neutral on the agreeableness scale, and a negative score would suggest you might not be the most agreeable person.

Why do psychologists complicate their lives? Why not simply ask people if they are trustworthy? The reason is that there is no right and wrong answer, and how you phrase the questions matters. It is fine to score negatively on the agreeableness scale since everybody is different, and not everybody is an agreeable person. This is the reason why the first question is phrased positively, and the second question is phrased negatively – or "reversed-scored" in psychology jargon. Longer big five questionnaires may include more questions about agreeableness, asking, for example, if you put the needs of others before your own, are empathetic, etc.

What about the other personality traits? **Conscientiousness** is about self-control and high levels of thoughtfulness. Conscientiousness also indicates self-discipline and the capability to regulate impulses. It is lower for young adults than for older adults (Noftle & Gust, 2019), the opposite of mind wandering propensity, which is higher for young adults. To test your conscientiousness, consider the following questions:

1. "I consider myself someone who 'does a thorough job'."
2. "I think I 'tend to be lazy'."

Say you respond that you neither agree nor disagree with the first statement (score three) and agree a little with the second one (score four). As with the agreeableness score, we subtract the second score from the first. Then your conscientiousness score would be three minus four equals minus one. Having a negative score is fine, and might mean your life has less structure, and you are happier with it.

The **extraversion** personality trait needs no introduction – it is about being energized by the company of others. The two statements are:

1. "I consider myself outgoing and sociable."
2. "I am a reserved person."

If you strongly agree with the first question (score of five) and strongly disagree with the second one (score of one), then you will obtain the highest extraversion score of five minus one equals four – a very extraverted person.

The last two traits are **openness** and **neuroticism**. As we will see in the next section, these two personality traits influence mind wandering the most. Openness is associated with having artistic interests, being interested in new experiences, meeting new people and experiencing new cultures. Less open-minded people like to focus on a few specific interests and tend to be aversive to change. Neuroticism is characterized by sadness, moodiness and emotional instability. You score high on the neuroticism scale if you tend to experience negative emotions.

To determine if someone is an open person or not, you would be asked to rate assertions about imagination and creativity. The questions below are extracted from the shorter big five personality questionnaire we have mentioned before.

1. "I consider myself as someone who has an active imagination."
2. "I consider myself as someone who has few artistic interests."

So to obtain a high score, you must think you have an active imagination (such as five on the first question) and have strong artistic interests (such as one on the second question). Then the difference between the score is highest (four in that case).

Now let's evaluate the **neuroticism** trait – also called emotional stability. Using the same scale as above, score each statement as truthfully as you can:

1. "I am someone who is relaxed and handles stress well."
2. "I am someone who gets nervous easily."

Write down your scores. The score is calculated by subtracting the score of the first question from the score of the second question.

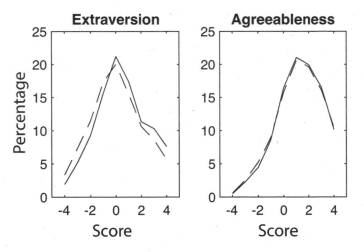

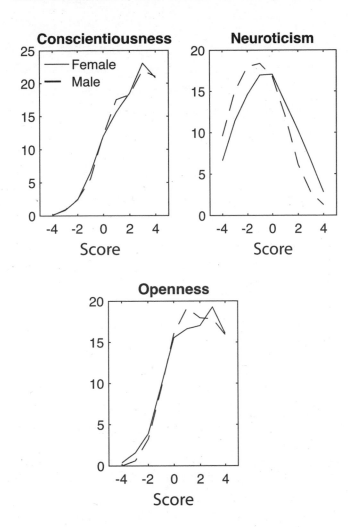

The figure above shows the big five personality scores for about 1,100 individuals participating in personal growth retreats (Wahbeh et al., 2022).

Although scores vary from -4 to 4, you will see that most people's scores are between -2 and 4 depending on the personality trait. By placing yourself on the curve for agreeableness, conscientiousness, extraversion, openness, and neuroticism, see how you compare to others in this population.

Your personality scores and mind wandering

Now that we've looked at how personality traits can be measured, I'll explain how personality traits influence mind wandering. Openness to new experiences is definitely a predictor of diminished mind wandering. It has been found that the more open people are, the less their minds wander (Baird et al., 2012).

Openness is also a predictor of increased daydream interest. On some tests, open-minded participants score high for questions such as, "I find my daydreams are worthwhile and interesting to me," or, "I imagine solving all my problems in my daydreams." This result might seem somewhat contradictory given that open-minded people mind wander less: how could these people's minds wander less when they enjoy daydreaming more? One consideration is that those scoring high on the openness scale engage in mind wandering and daydreaming when circumstances allow or when it is helpful but are very capable of concentrating when necessary. Or maybe those who are not as open try to suppress their thoughts more? Since suppressing our thoughts usually has the opposite effect, as we will see later, less open individuals might experience more spontaneous thoughts.

The type of thoughts your mind dwells on also differs depending on your personality scores. As you might expect, agreeableness is a predictor of pleasant thoughts. If you score high on the agreeableness scale, you will have more positive thoughts than those who score low. Perhaps you think more about past pleasant experiences, such as time spent with friends, and pleasurable future plans such as a planned vacation. Conscientiousness also leads people to more pleasant thoughts, although the association is weaker than agreeableness (Rusting & Larsen, 1998).

Neuroticism and mind wandering

The reverse is true for neuroticism. People scoring high for neuroticism have more negative thoughts and more thoughts about the past. They might brood over sad moments in their life or fear potential cataclysms. Depressed and anxious individuals also have more mind-wandering episodes about negative thoughts associated with past events (Joormann et al., 2012). It goes both ways. Individuals who experienced more mind-wandering episodes focused on negative events from the past reported lower levels of mood and happiness (Smallwood et al., 2009).

Extraversion and neuroticism both predict the presence of racing thoughts (Watson et al., 1994). When experiencing racing thoughts, each thought replaces the preceding one in rapid succession: "I need to go grocery shopping. Oh, then I need to get gas. I then need to go home and cook, etc." You are more likely to respond positively to the statement, "Right now, my thoughts are racing" if you score high on one of these two personality traits.

Mind wandering, memory and attention

If your personality affects when your mind wanders, what about your memory? Researchers have observed that higher memory capacity means decreased mind wandering. Memory capacity is your capability to remember concepts for a short time – so if I give you a sequence of 10 numbers, let you study them, and then ask you a few minutes later to recall them, on average will remember about seven. You would say that their memory capacity is seven items, and this number differs among individuals.

The higher your capacity at remembering things, as measured by standard working memory tests, the less you tend to mind wander (Kane et al., 2007). This intuitively makes sense because if you have better memory capacity, you can remember the task you are currently doing and do something else at the same time. In other words, you are less likely to forget the task you are doing. Other possible explanations are that people with high memory capacity also have higher IQs and an increased capacity to concentrate on solving problems. Nevertheless, the association between memory capacity and propensity to mind wander is relatively weak, which means that, although there is an association, people with high memory capacity might mind wander just a few percent less than people with low memory capacity.

Studies on memory capacity and mind wandering also support the idea that mind wandering depends on context. Those with high memory capacity can modulate mind wandering based on the task's difficulty (Soemer & Schiefele, 2020). In these experiments, the task involved remembering a set of objects and making decisions based on this set. On easy tasks, with few objects for which not much attention is needed, people with high memory capacity tended to have more mind-wandering episodes, probably because they can sometimes continue to do the task and mind wander simultaneously. Fewer mind-wandering episodes occur on more complex tasks requiring greater concentration, consistent with the idea that mind wandering depends on the context.

Attention and mind wandering

Another study shows the strong association between the rate of mind wandering – how many mind-wandering episodes people have – and our capacity to concentrate. There are different types of attention, and this result involves focused attention to visual objects. In this same study, researchers found that when people's minds wander, they are unable to inhibit their responses (Jana & Aron, 2022). For example, imagine a task where you are asked to press a button whenever a green dot appears on a computer screen. This is an easy task. However, sometimes the green dot is quickly followed by a red dot, and you must not respond – or stop yourself if you have initiated a movement to respond.

In daily life, a similar situation would occur with a green traffic light that transitions directly to red, not passing through orange. Imagine that you are driving and see the green traffic light. You start accelerating but it suddenly turns red, and you need to slam on the brake. More impulsive individuals and individuals scoring high on the neuroticism scale – as measured by questionnaires such as the big five we mentioned earlier – may respond slower at this driving test.

They might fail to stop accelerating and burn the red light. This is not because they are emotionally impulsive and more prone to wanting to jump a red light. Instead, their brain cannot detect the change of traffic light color in time to stop accelerating. Researchers showed that those failing to inhibit their reactions tend to have more mind-wandering episodes (Arabaci & Parris, 2018). One reason might be that they do not become aware as fast as others that they need to stop doing something – be it driving when the light turns red or stopping their movement in its course. As their mind starts to wander, they might not be able to prevent it from doing so.

Another type of attention linked to mind wandering is the capacity to focus visual attention amid distractors. Say, for example, trying to read this book when there are distracting elements in your environment. For example, you might be trying to read on the subway. Can you focus and ignore the activity in your environment, or do you get distracted easily? People with low spatial attention capacity tend to experience more mind-wandering episodes (Keulers & Jonkman, 2019). It intuitively makes sense. If you have difficulty concentrating, it might be because you get distracted easily, and your mind starts wandering.

Conclusion on cognition and mind wandering?

It makes sense that our capacity to focus attention to make quick decisions or ignore irrelevant visual content would be related to mind wandering. In both cases, increased control of our attention shows that we do not get easily distracted by external events or internal thoughts once we decide to do something.

The relationship with working memory is also sensical. If we can hold more in our memory, we are more likely to remember the task we are doing, so if mind wandering, we might realize it sooner. Of course, these are ad-hoc explanations, and the underlying cognitive mechanisms are more complex and not yet fully understood. Hopefully, more research will shed light on the relationship between cognition and mind wandering. Using what we have learned in this chapter, we will see later the techniques we can use to minimize mind wandering.

CHAPTER 9

WHY DO OUR MINDS WANDER?

Imagine a world where our mind never wanders, and you are concentrated on watching a riveting movie, totally immersed. Without your mind checking out sometimes, it is unclear if you would be able to pull yourself off the movie. Assuming the movie lasts forever, you might not think to feed yourself or to go to sleep. Most of us have experienced moments in our life when we were in "the zone," where three hours feels like one because we are so focused on something. Again, we might feel reluctant to stop our activity for any reason. In this case, mind wandering is useful to remind us to eat for example. The thought "I am hungry" sprouts out of our brain in the same way mind wandering does, and is likely based on a similar mechanism. (The thought "I am hungry" qualifies as mind wandering of the stimulus-dependent type since our hunger triggers it.)

Mind wandering is part of normal brain functioning

Mind wandering might have evolved to make sure that our mind does not get absorbed too much into one task, and that other thoughts or priorities in our lives are given a say. Scientists call this "attentional cycling" (Mooneyham & Schooler, 2013). Attentional cycling refers to the idea that the brain needs to periodically disengage from a task to assess its progress, evaluate whether it's still important, and consider whether other priorities or goals have emerged. By doing so, we can shift our focus to other tasks or goals requiring attention and ensure that we are allocating our mental resources effectively. The expression "your mind wanders because your brain whispers" captures that idea. When watching a movie, your mind will still wander from time to time about what to do next. Upon becoming aware of these thoughts, you might decide that it is time to stop and do something else.

In fact, the evolutionary advantage of mind wandering is supported by studies that compare mind wandering across different ages (Jackson & Balota, 2012) showing that younger individuals tend to mind wander more than older individuals. It is possible their mind is more active because they spend more time trying to figure out and interpret the world. For example, managing personal finances can be a difficult topic for children to learn, and it pays off if their mind drifts to that topic from time to time, to weigh the pros and cons on the best course of action. By contrast, older individuals do not need this additional mental activity. They have learned through trial and error plus decades of experience – another clue to the function of mind wandering.

Another indication that mind wandering is part of normal brain functioning is that, no matter how hard you try, it is usually impossible to completely suppress it. As a beginner or intermediate meditator, if you sit quietly focusing on your breath, you will spontaneously start thinking about a variety of topics. It could be your grocery list, your relationships, or the meditation you are doing at that moment. Thoughts will pop up even if you try very hard not to have them.

Even experienced meditators, with decades of meditation behind them can only stop mind wandering for a couple of hours at a time when they are experiencing samadhi, a state of intense concentration achieved through meditation. These mental states of clarity without mind wandering are rare, and experienced meditator minds eventually start wandering again.

Mind wandering and creativity

Mind wandering has also been associated with creative thinking. Scientists have conducted experiments showing that when our mind is allowed to wander, we tend to be more creative. Let's try this.

EXERCISE: MIND WANDERING FOR CREATIVITY

Scientists have designed creativity tasks, where you show people an object (such as a brick), and give them a couple of minutes to come up with as many uses as possible for that object.

1. Take a piece of blank paper and a pen and sit in a quiet place where you know you will not be disturbed.

2. Set a timer on your smartphone for three minutes.
3. Think of how you could use the object depicted opposite in a standard but also creative way. Come up with as many uses for the brick as possible.
4. Start now and then come back this page.

This is what I came up with in three minutes:

- Build houses
- Make a floor
- Make a retaining wall
- Paper weight
- Weight on a tarp
- Break window if you're stuck outside your car
- Stepping stool
- A hammer
- Sit on it when it is wet on the ground
- Use for weightlifting

The more uses you come up with, the higher your creativity at that moment.

Research has shown that, if allowed to do nothing for a few minutes after being shown the image, your performance is better than if asked to do a task that forces you to focus attention. In other words, if you are allowed to mind wander, you automatically come up with more creative answers (Baird et al., 2012), likely because your mind cannot stop solving problems. We have another sign of the purpose of mind wandering, as a mechanism by which our mind continues to anticipate the future even when doing a task – perhaps giving us an evolutionary advantage compared to other species. When our ancestors were chased by a predator in the savannah and needed to take refuge in a tree, mind wandering might have allowed us to find creative solutions to escape or have the upper hand in confrontations.

Mind wandering to help us remember

The formation of long-term memory is a complex process. It is generally accepted that there are two types of memory: explicit and implicit. Explicit memories – or declarative memories – correspond to our capacity to recall facts and events. It takes effort to retrieve these memories, like if I ask you what you did on your birthday last year. It will require some concentration on your part to recall that event. Try it! You will need to concentrate, sometimes even close your eyes. You might try to replace events in their context, and remember where you were at that time. Perhaps you could try to remember what kind of gifts you received. To do that, you may need to scan your memory for all the recent gifts received to find the one you got for your birthday. Sometimes, the memory eludes you and will be on the "tip of your tongue" for a little while before you can remember it.

Implicit memories correspond to learned skills, such as reading, driving or playing an instrument, and do not involve any conscious efforts. When we learned to drive, we had to pay close attention to all the objects on the road surrounding us. It would often feel overwhelming to have to pay attention simultaneously to what was on the road, anticipate other drivers' reactions, and also control your own vehicle acceleration and steering, and use the proper signal at the right time. This was overloading our senses. Yet, a couple of years later, driving will usually have become second nature. Your mind is on autopilot and you are able to navigate complex situations at traffic intersections without even noticing. In fact, you might be able to do other tasks at the same time, such as having a conversation. The activity of driving is then considered to be an implicit memory. When individuals engage in mind wandering during a task that relies heavily on implicit memory, such as driving or playing an instrument, their performance tends to remain relatively stable.

In contrast to explicit memory, we do not need to make any effort to access that type of memory. When reaching a traffic intersection with your vehicle, you need not search your mind for the different traffic laws as you would with explicit memory. Instead you know instinctively what to do. One other reason scientists believe these are two different types of memory is that amnesic patients may lose their explicit memory, but keep their implicit memories. They may fail to remember facts in their life, but are still able to read and drive.

Memory consolidation during sleep

Explicit memories, such as the events of our day, are stored during wake time in the hippocampus, a deep central brain structure. During sleep, these memories are "consolidated" – the process by which new information is transformed from a temporary, fragile state into a more permanent, stable form that can be retrieved later. Dreaming helps recall the day's events and removes memories that no longer need to be remembered. For example, you do not need to remember that you started the dishwasher before dinner a week ago.

A group of experiments found that a specific sequence of neurons was activated in the hippocampus when rats explored a maze, and these same neurons were activated in the same sequence at night. Researchers placed rats in mazes during the day and studied the activity of neurons in their brains – some neurons are only active when the rat reaches a specific location in a maze. As the rat explores the maze, the scientists follow where they are by looking at which neuron is active. At night, the same sequence of activation of neurons is observed: are the rats dreaming that they are exploring the maze? Consistent with the idea that sleep is necessary for memory consolidation, people with insomnia have reduced memory consolidation.

Microsleep

One popular hypothesis among scientists is that mind wandering is a form of microsleep, used by the brain to replay recent facts and events and increase our memory of them. When our minds wander, we might recall recent events or plan what we must do during the day. As such, it is an integral part of how the mind functions, and trying to prevent our minds from wandering could disrupt this system. If we could magically remove mind wandering, we might lose the ability to remember recently formed memories.

The brain uses sleep time to wash out some waste products (byproducts of cellular metabolism and other metabolic processes that accumulate in the brain tissue over time), as well as other substances like free radicals and inflammatory molecules. The wastes are removed by cerebrospinal fluid, a clear, colorless, watery fluid that flows in and around your brain. Sleep is critical for healthy cognition, and lack of sleep has been linked to increased waste products in the brain and various memory and cognitive deficits, including increased daydreaming (Marcusson-Clavertz et al., 2019). If we do not sleep enough, we might need to compensate with mind wandering during the day. One study found that during sleep, the glymphatic system – a network of vessels in the brain that removes waste products – becomes more active and efficient in clearing beta-amyloid from the brain (Xie et al., 2013). Another study found that periods of rest and relaxation during wakefulness, such as daydreaming, may also increase the activity of the glymphatic system and facilitate the clearance of waste products from the brain (Fultz et al., 2019).

Do animals' minds wander?

To find out the function of mind wandering, would it be possible to study and understand it in animals? Human daydreaming and mind wandering most likely emerged with language, so it is unclear if animals experience them. Research indicates that some animals could experience some forms of mind wandering. Animals are capable of mental time travel (Roberts & Feeney, 2009), meaning they have some form of long-term memory and may use this to direct their behavior. For example, chimpanzees in captivity have been shown catching projectiles to throw back at zoo visitors later. They have also been observed to hide these projectiles from their keepers, so they would not get caught. This behavior shows new and creative behavior that suggests planning, thinking and potentially creative mind wandering.

Remember the rat's brain neurons replaying their maze exploration during their sleep. As we saw earlier, dreams and daydreaming can be considered a form of mind wandering, so it is possible that these animals' minds wander. However, because it is easier to ask humans what they think than it is with animals, mind-wandering research will likely continue to rely on human studies.

Mind wandering and the ego

Most of the thoughts we have while our minds wander involve us as an actor and a story of us: "So and so did that to me, and this is unacceptable. I must find a way to make it stop." This is the ego. As we will see later in the section on trying to deal with our wandering mind, any story we tell ourselves repetitively is like a virus that has hijacked our brains. These viruses may maintain themselves by replaying the story like a broken record during mind wandering. This is also supported by neuroscience: thoughts involve the activation of neural pathways in the brain. As you activate these pathways, they strengthen. Let's use the analogy of a country dirt track – if nobody travels on that path, soon nature will reclaim it. However, if cars travel that way every day, eventually someone will decide to pave it and it becomes a highway. Neural pathways are similar – the more you rehearse a thought, the more you will think about it in the future. Ego-related thoughts might sprout once or twice by chance, but if we pay attention to them, they become self-sustaining. Mind wandering is a way for the ego thoughts to sustain themselves.

The modern interpretation of the ego started as a Freudian concept – the organized part of the personality, primarily responsible for planning and associated with reason and common sense. In this tradition, the ego is the judgmental part of us. However, it does not necessarily mean that it is the part of us that judges. One may consider a bottle half full, and the ego might not be involved. The ego is the part of us that performs judgment related to ourselves. For example, "The bottle is half full. Someone must have drunk half of it. This is mean. I am thirsty. Why were they not thinking of me?" is an ego-laden judgment.

The ego is problematic for most Eastern spiritual traditions and hinders spiritual development; one of the goals of most Eastern spiritual practitioners is to rid themselves of the ego. It is relevant for mind wandering because its function might be to sustain the ego, and we will see later how meditation affects mind wandering. Hopefully, there will be more direct research on egoism and mind wandering in the future to demonstrate the link between the two.

Why the ego evolved

One may easily imagine why the ego evolved, and by extension mind wandering that supports it. In a competitive world with limited resources, judgmental thoughts leading to the suppression of less capable competitors may become an advantage. For example, "I have to be better than everyone else to succeed," "I can't let anyone outdo me," "If I don't win, I'm a failure." These thoughts may help one acquire more resources and pass on genes to future generations. Without these thoughts, we might not aspire to achieve more, and society as we know it might not have developed.

In Eastern spiritual and cultural traditions, the ego is a synonym for separation. In these traditions, everything is tightly connected, including nature, space, time, and consciousness. Ego-related thoughts often negate that connection and present a perspective only valid from the narrator's standpoint. This ego concept is related to egoism. An egotistical person will first think of themselves. Self-interest comes first; the underlying ideology is that we need not think of others. It is also related to the hedonic treadmill. In the hedonic treadmill, our ego seeks pleasure, is satiated for a while, and then seeks more pleasure. The ego convinces us that if we obtain this object or attain this position in our career, we will finally be happy. This thought is a lie because as we eventually reach our goal, we realize another goal has replaced it. We think one million dollars would make us blissfully happy, but, if we happen to reach that goal, then we now need 10 million dollars to be happy. These ego thoughts relate to daydreaming, the spontaneous mind wandering mentioned at the beginning of this book. We are daydreaming about all the things we could have, and once we get them, we continue to daydream about new things.

The ego takes a bow

Ego thoughts and the associated daydreams or stories we tell ourselves while mind wandering are often unnecessary. For example, we may be practicing a sport for the first time. Sensation, concentration and learning take place. If we're trying surfing, we may think, "Ah, this is the way to stand on the board," or, "If I lean when standing, I will fall." These thoughts do not involve the ego. If, after a successful run, we spontaneously think, "I did it." The I is the ego, taking credit. Spiritual teacher Rupert Spira compares the thought "I did it" to the clown at the circus coming on to take a bow for other performers' accomplishments. Clearly this spontaneous thought was unnecessary, and we could have dwelled in silent satisfaction. The thought "I did it" serves no purpose.

This example was about a positive thought but a negative thought could also be an ego thought. Imagine you are experiencing chronic back pain. The pain is unpleasant, but there is also the spontaneous narration about that pain such as, "I will never be able to do sport as I used to. I feel miserable." These spontaneous thoughts make us unhappy and depressed. They are what mindfulness teacher and researcher John Kabat-Zinn calls the "second arrow of pain" – the first arrow being the pain itself. We cannot avoid the pain, but the subsequent thought about the pain that spontaneously arises seems unnecessary. It amplifies the pain we feel. This type of mind wandering can be controlled, and intervention methods have been shown to decrease pain.

Here, the delineation between intentional thought and mind wandering becomes blurry, and it is unclear which thoughts are voluntary. Do you remember how we described mind-wandering thoughts at the beginning of the book: some are spontaneous, but voluntary. Could it be that ego thoughts – when we think about ourselves – sprout into consciousness as mind wandering would, and are mind-wandering thoughts? The thoughts when our mind wanders are undoubtedly involuntary – most often, we are trying to concentrate and not have such thoughts. However, this could also be the case for thoughts we think are voluntary such as when we think about ourselves. We mentioned the virus analogy before where thoughts pop up and disappear, and the more we give them attention, the stronger they become. Ego-related thoughts could pop up when our mind wanders or during a sequence of voluntary thoughts. Can we be sure that voluntary thoughts are willed to exist by us? We will come back to this topic later.

Ego thoughts make us unhappy

As spontaneous ego thought takes more and more of our mental space, unhappiness creeps in. Judgmental thoughts, which are often spontaneous and thus a form of mind wandering, make us unhappy because judgment creates a barrier between ourselves and the world. As we have seen before, we are most happy when we are embedded in the task we do (Killingsworth & Gilbert, 2010), unencumbered by parasite thoughts. Negative judgments triggered by our environment – another type of mind wandering we mentioned earlier – like "I do not like the weather today!" put us in a bad mood. Positive judgments are often laden with envy or jealousy, "I like this person's outfit and the way they look. . ." with the implicit thought, "I wish I looked that good." Or it could be a judgment that feels positive, "I did so much better than that other person," that makes us happy. However, behind this self-gratifying judgment is also the fear of failing in the future and being diminished, so both positive and negative spontaneous ego thoughts parasitize our mind and make us unhappy.

Ego and the fear of death

Spontaneous ego-centered mind wandering also makes us unhappy because it revolves around our biggest fears – of impermanence and death. Why? Because not only does their content refer to some impermanent object, but also because the thought itself is impermanent. It comes and goes. This is very important. All mind wandering thought comes and goes and, when it contains an ego-centered thought, as the thought vanishes, we experience a little death. As the thought disappears, we feel our impermanence. The content of the thought also refers to some impermanent object. Even if we are religious, when we give our attention to the positive ego thoughts (for example, "I look so good,") we also implicitly know that this "I" might not always look so good in the future. As we grow old, our appearance changes, usually not for the better. This "I" the thought refers to (our body which looks so good) will die one day.

Experienced meditators will tell you that if you search for this "I" within yourself, you will not be able to find it. This "I" is a concept made tangible by the impermanent thought. However, it is not real. There are meditation practices where you only ponder one question, "Who am I?" You do so for years for several hours a day. One thing you will realize when you dig in your mind that way is that you are not your thoughts. Thoughts and concepts come and go, and you do not. The "I" in the "I look so good" concept is not who you are. Therefore, it does not exist. Ego thoughts know they are about a non-existing entity and resist this with all their strength. Like viruses, they want to grab your attention to continue to exist as thoughts, polluting your mental space and generating stress. Judgmental thoughts lead to unhappiness, and judgmental thoughts are rooted in the spontaneous ego-centered thought occurring during mind wandering.

Enlightenment

Eastern spiritual tradition argues that it is possible to decrease the occurrence of mind wandering and ego-centered thoughts. The ultimate goal of most of these traditions is to reach enlightenment, a state where ego-centered mind wandering cannot sway you. When enlightened, you may still think and act in the world. You may still have opinions and preferences. Still, you are not taking your ego-centered mind wandering seriously, and you can be at peace and fulfilled even if proven wrong. Say you start to mind wander and the thought pops up into your mind, "I am growing old." Then another thought may pop in your mind, "This is just a thought, I can see and feel that I am more than my thoughts and that it is just passing through my mind," and then you might just rest in the present, content to feel alive, without thought, and without the idea of growing old having any grasp on you and your well-being. After some time, the thought "I am growing old" does not even bother popping up.

I am not enlightened so it is not easy for me to talk about enlightenment. However, it is generally accepted that enlightenment is more of a lifelong path than a goal. A goal is only for the ego on the hedonic treadmill. Enlightened individuals tend to surf the wave of consciousness wherever it takes them and no longer have self-centered goals. In doing so, they reach peace and happiness. Much of the next part of the book involves techniques inspired by Eastern tradition and adapted to the West. Some of the practices we will present will allow you to work on some of your ego-centered mind-wandering thoughts and decrease their grasp.

PART 3
TAMING YOUR MIND

CHAPTER 10

DEALING WITH PERVASIVE THOUGHTS

People tend to want to change their behavior without changing themselves. They are scared of the change for good reasons: if they truly want change, they will become a different person. Many people are not ready for that – they say they are but they cling to their identity as if their life depended on it. If we want to tame our minds, though, we must be ready to question ourselves. Thoughts are not objects we can dispose of. They are like organisms living within us, and if we let them go – even negative thoughts – we may feel like a part of us is dying.

Are you ready?

Most people think about working on their mind wandering only if it starts to make their life very difficult – at the beginning of the book we heard about Bobby with his concentration issues. The wandering mind is like breathing – if you are breathing fine, there is no need to change your breathing pattern. If you have asthma or a breathing problem because of anxiety, it is time to work on the issue. Remember Jane with her persistent thoughts about her political beliefs? She unconsciously clings to the thoughts because they are part of who she believes herself to be, and ceasing to believe them would mean the death of her identity. If she is desperate and genuinely ready to change, she is finally ready to work on herself and her negative mind-wandering patterns. For this reason, the techniques I present in this section also help deal with negative thoughts we want to let go of. It takes effort to curb mind wandering so the motivation must be strong.

This book will not provide you with a miracle method. It will give you the tools and support to change yourself if you are ready. I have applied all these tools myself and can testify to their efficacy. I have not reached the pinnacle of inner peace, but I believe I am much better off than when I started. It is a lifelong journey.

The mind as a tool

In previous sections, we have seen that the mind is a tool to study mind wandering. We asked people about their thoughts and feelings and made conclusions about the function and dynamics of mind wandering.

We must tame our minds and teach ourselves to detect and analyze mind-wandering episodes. Mind training and mind taming are central themes in the approaches that follow.

Mind wandering as a virus

We mentioned earlier that mind wandering might be supported by virus-like processes in our brain and mind. For example, say you have a pervasive thought, "I want to go on a vacation to Hawaii on a boat," and every time you encounter something in your daily life that reminds you of that, be it in the media or talking to a friend, you will think of your desire to go to Hawaii on a boat. For example, you might have that thought every time you hear the words ocean, boat, island, etc. When your mind wanders, you also daydream about that vacation. This thought, although positive, is hijacking your mind and, to some extent, making you unhappy (the thought tells you that you are not happy now and will only be once you take that vacation). Once you have realized that dream, these words will no longer trigger that thought.

The example above was generally positive, but the same could happen with a negative thought, such as, "I feel I am overweight because I eat too much ice cream like my uncle." The trigger words will be uncle, eat, ice cream and overweight. As shown in the diagram below, these words are associated with my body perception, so thinking about myself or how others perceive me will trigger the thought. Whenever I think of myself, my uncle, how I look, or about eating, the negative ice cream thought pops into my mind.

This vicious cycle is purely mental but can also become physical. The thought is stressful for us and could lead to anxiety and then to emotional eating of ice cream. The diagram below shows a mechanism where the thought is coupled with behavior. We think the stressful thought. We eat to suppress negative emotions. Consequently, we gain weight and have more negative eating thoughts. This shows how the thought can sustain itself through our behavior and our body.

In the brain, these thoughts are supported by sustained patterns of brain activity. For these word/concept associations to be stable, they need strong connections. If one node (group or area of neurons) is activated, the rest of the related network becomes activated. For example, if the node "uncle" is connected to the node "ice cream" and "overweight" and the node "me" then any activation of one of the nodes will lead to the activation of the others. This association is supported by strengthening connectivity between multiple brain areas (memory, planning and emotion). Of course, this is an oversimplification, as millions of neurons in disparate brain areas may be involved in sustaining these thoughts, so millions of connections get strengthened. However, the picture is clear. The more you think about a negative thought, the more it casts a web of connections in your brain and mind to stabilize itself.

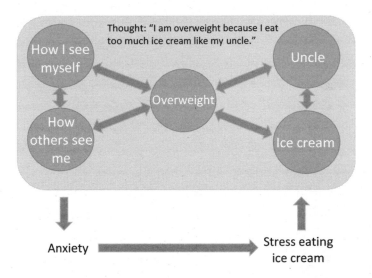

How do we get rid of these thoughts?

The figure above shows the dynamic of a pervasive negative thought. So, what should we do if we wish to change this? Imagine I ask you not to think of a pink elephant for five minutes. Whether you want to or not, you will think about the pink elephant (or the fact that you should not think about it) multiple times over the five minutes. All thoughts are like the pink elephant, we cannot simply wish them to disappear. If we do that, we must think about them: to stop thinking about the pink elephant, I must think about it which defeats the purpose. The more we do not want to think about the pink elephant, the more we do. One way to succeed is to "reappraise" the thought, understanding that these thoughts are either false or that they are hurting us and our bodies. We know the thought is bad for us on some level, but we have not realized how bad it is – this realization must come as an "aha" moment. Only then can neural mechanisms take place to suppress the activity because it is no longer supported by one of its nodes (the node "I genuinely believe this" or the node "this is part of who I am"). Usually, it takes dedicated practice to decrease the grasp of a negative thought to an acceptable level.

EXERCISE: REAPPRAISING A THOUGHT

The first step to reappraising a thought is to become fully aware of it. To do so, follow these steps:

1. Take a small piece of paper, and write down the pervasive thought that is upsetting you. For example, "My partner is not considering me."
2. Now place the paper and a pen in your pocket and for the rest of the day, write down a bar every time your mind wanders onto the pervasive thought.
3. When doing this, do not count mind wandering that occurs within five minutes of the previous count. Also count the thoughts you have voluntarily, for example, if your partner says something, and you have the, "My partner is not considering me," thought in response to this interaction.

If possible, continue the exercise over several days. Even within one day, though, some change may happen: our mind is lazy and prefers not doing exercise of that sort. Your mind might decide it is less effort to not have the thought than to have to write it down. Even though you cannot decide not to have the thought – do not think of the pink elephant – giving yourself extra work when you have the thought could be enough to have it decrease in intensity and frequency.

The other key result of this exercise is that you realize how much you have the thought and the time you are wasting in having it. You also understand that the thought is not part of you, because you do not want it. This supports your mind to simply decide to let go of the thought if it comes with a conscious realization that this is not you: if the thought pops up in your mind, the counteracting thought will pop up, "Not again, go away, thank you." It is no longer part of you and you do not identify with it. It has lost its grip.

CHAPTER 11

COGNITIVE RESTRUCTURING METHODS

Another technique to train our mind is to reframe our thoughts using Cognitive Behavioral Therapy techniques, CBT for short. There are numerous ways that thoughts upset us and cause problems, such as black-and-white thinking, catastrophizing, overgeneralizing or personalizing.

These are called cognitive distortions. In black-and-white thinking, scenarios or individuals in our entourage are either right or wrong, and there is no middle ground. Polarization in politics is a good example, where we demonize the opposite camp. The same happens in intimate relationships, where we are convinced we are right and the other is wrong. Black-and-white thinking leads to "dehumanization," where we treat others as problems to solve rather than human beings to work with.

Catastrophizing is when we blow a small negative event out of proportion. No, you do not need to move because your neighbor is rude when you only see them once a week taking out the trash. Overgeneralizing happens when we generalize individual events: "You're never ready on time, so we are always late for everything." While accurate sometimes, this statement is unlikely to be true all of the time. Personalization is when you blame yourself for circumstances beyond your control. For example, an unexpected flood ruins your vegetable garden, and you think, "It is my fault; if I had picked all my vegetables last week, we would not have wasted so much." These thoughts serve a purpose: they bring us short term comfort, by reducing uncertainty (if I move, I will not have to deal with my neighbor), or boosting our ego by making us feel right. However, they also contribute to the vicious cycle of sustaining themselves and making us unhappy.

We all suffer from these to some extent, and our thoughts voluntarily and involuntarily reflect these biases. Cognitive restructuring techniques try to make you see the situation from a different angle.

Cognitive behavioral therapy

The way this works is through your therapist giving you a set of instructions to enable you to analyze your thoughts. For example, upon realizing you have a negative thought about someone or something, you would ask yourself:

- Is this thought based on emotion or facts?
- If it is based on facts, how accurate is this fact?
- If it is true, what is the worst that could happen?
- Is this a black-and-white situation, or are there shades of gray?

Let's take the example of blaming your partner for never considering you. You would first look for facts and might remember that episode when your partner decided to spend time with their friend instead of you. You would then try to challenge your perspective on the event and assess if alternate interpretations are possible. What counter-example might you have of opposite behavior?

EXERCISE: COGNITIVE RESTRUCTURING

A common exercise in CBT is "Cognitive Restructuring," a technique which has been shown to work in clinical trials. It involves recognizing and challenging negative or irrational thoughts (referred to as "cognitive distortions") and replacing them with more balanced and accurate thinking.

For example, a cognitive distortion might be if a person is feeling anxious about public speaking thinking, "I'm going to mess up and everyone will judge me." You can work on this with alternative, more realistic thoughts, such as, "I have prepared well, and even if I make a few mistakes, it won't be the end of the world."

Let's try:

1. Get a pen and paper and find a quiet place where you won't be disturbed for about five minutes.
2. Pick a thought that generates anxiety in you and that you often experience spontaneously and write it down on the piece of paper.
3. Write at least one sentence explaining why this is black-and-white thinking/catastrophizing/overthinking, etc.: the sentence should start with "This is a black-and-white/catastrophizing/overthinking thought because . . ."
4. Now write down at least two sentences of realistic nuanced thoughts that challenge the initial anxiety-inducing thought and instead provide a realistic assessment.

For me, my mind often wanders on the subject of how much I have to do and that I cannot keep up with the volume of work, which generates anxiety. This is a black-and-white thought because if I cannot do all the work in time, it will simply be partially done and/or delayed, so this is not an all-or-nothing situation. One example of dealing with this might be a time when I have too much to do – perhaps I promised to finish the second draft of this book by the end of the week but I also have to write and submit three grants within the next two weeks to pay my researcher salary, which doesn't feel possible. A more nuanced thought could be that the book is almost finished, that the grants are partially written, and that I have almost never missed a deadline. Another more nuanced thought would be that if I failed to meet the deadline, it would not be the end of the world. No life is at stake. There would be other grant opportunities, and I am sure my editor would grant me an additional week or two to finish writing the book.

ACCEPTANCE COMMITMENT THERAPY (ACT)

A closely related and equally effective technique is Acceptance and Commitment Therapy, or ACT for short. This technique works on restructuring thoughts, focusing on acceptance.

In one of the examples above, where you feel unappreciated by your partner, you would focus on why you desperately need to feel valued. Do you need to be appreciated to be happy? Happiness must come from within. When you realize that you do not need the other person's acceptance, maybe you will start enjoying their company more, and in turn, they will appreciate you more in return. This does not mean you must stay in an abusive relationship. You may decide that this relationship is no longer for you. However, if you move on, it will not be because of a lack of appreciation from your partner. Otherwise, you will, in all likelihood, fall into the same relationship pattern with someone else.

For more information on CBT, refer to one of the CBT workbooks (Gillihan, 2018).

EXERCISE: ACCEPTANCE COMMITMENT THERAPY (ACT)

Let's try a simple exercise in Acceptance and Commitment Therapy (ACT):

1. Get a pen and a piece of paper and find a quiet place where you won't be disturbed.
2. Identify a difficult or distressing experience or thought that you've been avoiding. This could be anxiety, fear or a painful memory. Write down the thought or experience in one sentence.
3. Now close your eyes. Imagine holding the thought or experience like a balloon. You see the balloon in your mind and it contains the thought.
4. Imagine letting it go, and it floating away into the sky. Visualize the balloon in the sky. Observe the thought or experience from a distance, as if you are watching it from the outside.
5. Ask yourself, "What would I like to focus on instead of this thought or experience?" Consider what values, goals or meaningful activities would bring you more happiness, fulfillment and well-being.

6. Write down three actions you can take in the next 24 hours that align with your values and goals.

7. Finally, let go of the thought or experience, and commit to taking these actions in the present moment.

This exercise can help you practice accepting difficult spontaneous thoughts and feelings, and redirecting your attention toward things that are important and meaningful to you. For more information, refer to the ACT handbook (Harris et al., 2019).

CHAPTER 12

USING THE THOUGHT DIAGRAM

My friend and colleague Antoine Lutz and his collaborators created a diagram to represent the inner space and how to navigate this space. In the simplified version below, mental states are organized along two axes: meta-awareness and reification. What are these? Meta-awareness is your awareness of your thinking process. During a conflict, a person with elevated meta-awareness will be aware of their thoughts and feelings and might be able to observe them and not act on them. By contrast, a person with low meta-awareness will ride the thoughts and emotions and embed them no matter how destructive they might be. They will believe the thoughts as they are having them and only realize later that they might have overreacted.

Reification corresponds to whether you believe objects are real or not – objects here meaning physical objects, but also people and mental objects. Our level of reification is also our level of belief that our thoughts are real. The more we believe our thoughts and ride them, the higher the reification level.

Mind wandering in the Thought Diagram

In the diagram, involuntary mind wandering is a low meta-awareness mental state since we are often unaware we are mind wandering. Suppose we were aware we were involuntarily mind wandering. In that case, we might decide to continue mind wandering (and it becomes voluntary mind wandering) or decide to resume focusing on the activity in which we are engaged. In both scenarios, becoming aware of involuntarily mind wandering interrupts it.

Involuntary mind wandering is also considered high in reification – because when our mind wanders, we believe our thoughts. We do not only believe them but embed them: we are them. There is also little critical thinking when our minds wander: we are not thinking deeply about issues, only scratching the surface.

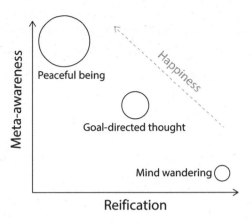

Voluntary thoughts

Goal-directed thoughts have a medium level of meta-awareness. When mentally solving problems in a work-related context, for example, we are aware of what we are doing. We can also voluntarily stop that activity. By contrast, when our minds wander, we cannot stop the thoughts since we are unaware of them. With goal-directed thoughts, our level of identification with our thoughts and objects is also generally lower. Our thoughts are a means to an end and do not define us. Voluntarily making our grocery list consists in accessing our memory, making mental calculations of how many of each item we might need, etc. It has a different level of intentionality compared to when we are mind wandering about what we may want to eat tomorrow and remembering that we must not forget to go and buy butter – and that it is frustrating that we forgot to buy it earlier. When consciously drafting a grocery list, butter is an abstract item. When mind wandering, however, it becomes the center of our thoughts and preoccupation. In the case of voluntary thoughts, our thinking is systematic and organized. In the case of mind wandering, it is disorganized and emotionally laden. Even though we are thinking about the same things, the level of awareness and de-reification is usually lower than when mind wandering.

To explain more clearly, I could have had the thought, which is likely common to a lot of people, that my partner should change to be the way I want her to be. It would definitely fall in the mind-wandering circle. This is a thought that I have spontaneously and that has high reification (I believe it is true). These thoughts do relate to my behavior, because I try to change my partner and she may resist – in the same way that I would resist if someone tried to change me – which could lead to conflict. Now, next time I have this thought, I will try not to fully engage in it, and will question the thought rather than believe it blindly. This will decrease the level of reification of the thought, and hopefully in time, I may realize that I do not need to change my partner. Instead I might need to change myself so I can become a more accepting person and accept my partner as she truly is.

EXERCISE: AWARENESS OF THOUGHTS, EMOTIONS AND BEHAVIORS

This exercise could be useful in helping you become more aware of your spontaneous mental states and to develop a greater understanding of the relationship between your thoughts, emotions and behaviors.

1. Draw a two-dimensional graph, like the one on page 159, with meta-awareness on the X-axis and reification on the Y-axis.

2. Take a few minutes to reflect on one recent spontaneous thought, and identify where it might fall on the graph. For example, you might have had a thought that was highly reifying (you fully believed it to be true), but with low meta-awareness (you weren't aware that you were having the thought).

3. Once you've plotted your thought on the graph, consider how it might relate to your actions and behaviors. For example, you might notice that thoughts with high reification are more likely to lead to impulsive or reactive behaviors, while thoughts with

high meta-awareness are more likely to lead to more deliberate and thoughtful responses.

4. Practice shifting your thoughts and emotions from one quadrant to another. For example, if you notice that you often have highly reifying thoughts, try to bring more meta-awareness to your thinking process. You could do this by simply noticing your thoughts and feelings as they arise, and then observing them from a distance, rather than fully engaging with them.

5. Repeat this exercise regularly to gain a greater understanding of the patterns in your thinking, and to develop skills in managing your thoughts and emotions.

This exercise can be a powerful tool for developing greater awareness of your mental states, and the ability to manage them more effectively.

Flow state is love

You may have noticed a third state on the diagram above labeled "Peaceful being" measuring low on reification and high on meta-awareness. In this mental state, we see our thoughts passing by but they do not disturb us. We may have negative thoughts, however, we see them as a train passing: we are aware of their transient existence and we are able not to jump on the train (meditators and those experienced in mindfulness will recognize this approach).

It isn't easy to describe that mental state for people who have not experienced it. The closest is probably being deeply in love. By love, I am not talking about the sustained love we may have for our family and friends, but the transient first love where we believe for a few weeks that we have found someone or something that completes us, and we are blissfully happy. In this state, we may learn disturbing news, for example, that our house was ransacked. We might be upset, but it will not affect our peace. We are fulfilled in the present moment, and adverse events and associated thoughts do not profoundly affect us. This state of "peaceful being" is similar to being in love, except we are in love with the world and feeling alive.

In this flow state of love, our level of reification is low. We are aware of the impermanence of objects and thoughts. Not only do negative thoughts leave us unperturbed, but so do positive thoughts. If $10,000 falls on our lap for one reason or another then our peace is as unchanged as if we had just lost the same amount, because this short-term gain or loss does not compare with the happiness we are experiencing. It is the realization that these events and thoughts are irrelevant to our mental state of deep love for the world. This sounds fantastic, but how do we get there? We get there by working on ourselves using the exercises in this book and others. It is a long journey.

What about content?

On the diagram, the size of the circles represents the aperture of consciousness, in other words, how much we are aware. This is similar to meta-awareness but different because it deals with our mind's content and not so much with our attitude toward it (although the two are certainly related). When mind wandering, the content of our consciousness is narrow. We are lost in our thoughts and usually can only be distracted by strong environmental events. When having goal-oriented thoughts, we might still be able to enjoy a piece of background music, so our consciousness aperture is larger. Finally, our consciousness content is expanded in the "peaceful being" state because we do not ride our thoughts, and we might even be able to witness the transition between thoughts.

I did not put rumination in this diagram. Ruminations are a form of contrived mind wandering, where we rehearse the same thought repeatedly. Rumination is often associated with depression or anxiety, and ruminative thoughts are usually involuntary. However, even as we become aware of them, we might be unable to stop them. These negative thoughts define who we are. Since our identity is tied to the thought, these thoughts' reification might be higher than regular mind wandering: we truly believe them. Paradoxically, meta-awareness could be higher than during mind wandering because we are often aware we are having these thoughts and do not consider them as involuntary as mind wandering. The exercises presented in this chapter may help us deal with such thoughts.

* * *

Some of the practices presented here are trying to move you away from a place of high reification and low meta-awareness to a place of low reification (de-identification with objects) and high meta-awareness (or high level of mindfulness). Hopefully, this leads to happiness, but as mentioned earlier, you cannot simply apply the techniques. The process of de-identifying with objects means your relationship with your belongings, family, job, preferred food and hobbies will change. Your conscious mind might want the change, but your body and unconscious mind will resist it with all their strength. The following sections will help us deal with resistance to change.

CHAPTER 13

MEDITATION TO CURB MIND WANDERING

In this chapter we talk about meditation and mind wandering – the wandering mind is engrained in most meditation practices, and even central to some. I started meditating in the Soto Zen tradition, where one is instructed to meditate on riddles called koans, such as, "Two hands clap, and there is a sound. What is the sound of one hand clapping?" or, "What is your original face before your mother and father were born?" The meditator ponders the riddle until they get an aha moment. After sometimes hundreds of hours of meditation, they realize the answer (which always has to do with the non-duality of all things). The answer is not important. There might be several answers, and the most important is that you get the aha moment, not the answer. When I did this practice, I realized that I was only working on the koan about a quarter of the time. My mind wandered the rest of the time, thinking of random topics. No matter how hard I tried, I could not prevent these thoughts from occurring. This became frustrating. The harder I tried and became irritated, the worse the mind wandering was. I thought I was a terrible meditator, only to learn later that this is the nature of our mind. Only during day-long retreats would my mind sometimes calm down.

Beyond meditation, what we talk about here would also apply to any type of sustained task where we need to concentrate. For some reason, I like to think of nuclear reactor surveillance staff. They must maintain attention at all times and monitor any problem with the reactor. Their job requires high concentration, yet it is repetitive, so their minds must wander. Because their minds fail to obey them, they might be more aware of mind-wandering phenomena than others – similar to a meditator trying to concentrate on their breath, for example.

The awareness that our minds wander and do not follow our will might also come during episodes of depression. Ruminative thoughts keep occupying our minds, even though we wish these thoughts would go away.

Concentrative meditation

Over millennia, Eastern traditions have aimed at refining methods to study the mind. There are many meditation traditions. Focused-attention meditation practices cultivate enhanced concentration and single-pointed focus on a given object in addition to developing meta-awareness. Mantra meditation practices generally focus on reciting a mantra, a single word or a sentence. As the meditator focuses on their mantra, they become hyper-aware of when their mind wanders. Other types of concentrative meditation revolve around non-duality. Non-dual practices are based on the idea that all things in the universe are interconnected and inseparable, and that ultimately there is no distinction between the self and the world. Loving-kindness and compassion meditation is yet another meditation concentration practice that involves cultivating compassion and usually involves mental imagery techniques.

It would be wrong to think Western Europe was immune to these practices during the past millennium. In the introduction, I mentioned the mystic 14th-century Christian text *The Cloud of Unknowing* which referenced mind wandering during prayer. Christian devotional prayer, the Kabbalah and Sufism are all types of Western practices in the Christian, Jewish and Muslim traditions which have many commonalities with Eastern meditation practices. Scholars often prefer to use Eastern practices because they can be practiced even if one does not believe or adhere to the underlying religious narrative. They depict a systematic approach to studying the mind, closer to the Western psychology secular approach.

EXERCISE: "JUST SITTING" MEDITATION

Open-monitoring meditation is a meditation where the person tries to be aware of their external and internal spaces. In Zen, this is called Zazen meditation, often translated as "just sitting." Imagine sitting with your eyes closed or almost closed, and you hear a bird singing outside. The bird song becomes the object of your meditation. Imagine now that you start thinking about how beautiful the song is and how good you feel. The thought of feeling good and the associated sensation become the object of your meditation. There is no distraction possible in open-monitoring meditation, as the distraction would at once become the object of meditation.

During open-monitoring meditation, you may sometimes experience your spontaneous thoughts from a distance. You see them pass, but they do not become you. It is an exhilarating feeling to realize that you are not your thoughts. This type of meditation requires practice, and even experienced meditators have not perfected it.

Here's a five-minute guided open-monitoring Zen meditation practice:

1. Set a timer on your smartphone for five minutes.
2. Find a quiet, comfortable place to sit with a straight back and close your eyes. Take a deep breath through your nose, hold for a few seconds, and exhale slowly through your mouth.
3. Focus on your breath, feeling each inhale and exhale.
4. Allow your mind to be open and receptive to any thoughts, feelings or sensations that arise. Observe them without judgment, simply acknowledging them as they come and go.
5. If your mind starts to wander, simply bring the focus back to your breath and the sensation of breathing.
6. Continue this practice for the next five minutes, observing any thoughts or sensations that arise and bringing your focus

back to your breath whenever your mind starts to wander.

7. When the five minutes are up, take a deep breath in and out, and slowly open your eyes. Take a moment to feel the sense of presence and awareness that you have cultivated within yourself.

Even though "just sitting" sounds like the easiest thing to do (and it is), it is challenging because your mind has nothing to do, and so tends to go all over the place. This meditation should be called "just noticing" rather than "just sitting" because your task is to notice more than it is to sit. In this meditation, distraction does not exist, because being aware of the distraction is the focus of the meditation.

Mindfulness

The meditation depicted above is an ancestral practice, but there are also modern secular techniques. In 1979, Jon Kabat-Zinn began to study the effect of mindfulness. Kabat-Zinn was a student of Buddhist meditation and a scholar at MIT. He studied the influence of a combination of meditation traditions on a group of patients with chronic conditions who did not respond to drug treatment. His method was named mindfulness. For Kabat-Zinn, mindfulness is the "awareness that arises [. . .] in the present moment, non-judgmentally [. . .] it's about knowing what is on your mind." He developed an eight-week training protocol where participants are asked to practice for a short period briefly every day, culminating in a "retreat" at the end, a several-hour silent meditation practice. These programs work to treat depression and reduce pain (Kabat-Zinn, 1990) and are called Mindfulness-Based Stress Reduction and Mindfulness-Based Cognitive Therapy; it is now used in thousands of hospitals worldwide. If you go through one of these programs, first you are asked to be mindful about eating a raisin. Then you practice a different type of breath-centering meditation. These methods are inspired by, and sometimes carbon copies of, Eastern meditation practices. Although the therapy group only meets once a week, students are asked to practice meditation over eight weeks for 45 to 60 minutes daily.

These techniques have been proven to decrease the occurrence of mind wandering. In a key study, researchers asked students to participate in a mindfulness intervention and observed that people reported fewer task-unrelated thoughts after the two-week training period. We will come back to that study later (Mrazek et al., 2013).

EXERCISE: RAISIN MINDFULNESS

Let's try the well-known five-minute mindfulness practice on awareness of a raisin:

1. Get a handful of raisins, place them on a plate, and set a timer on your smartphone for five minutes. Find a quiet, comfortable place to sit and close your eyes.
2. Take a deep breath in through your nose, hold for a few seconds, and exhale slowly through your mouth.
3. Keeping your eyes closed, take one raisin and place it in the palm of your hand.
4. Now open your eyes and focus your attention on the raisin, observing its shape, color, and texture.
5. Take a moment to examine the raisin closely, noticing any details or nuances that you may have previously overlooked.
6. Pick up the raisin with your thumb and forefinger and bring it close to your nose. Take a deep breath in, and notice the scent of the raisin.
7. Place the raisin in your mouth and pay attention to the sensation of the raisin as it touches your lips, tongue and roof of your mouth.

8. Chew the raisin slowly and deliberately, paying attention to the texture and taste. Take note of any changes in flavor as you continue to chew.
9. When you have fully chewed and swallowed the raisin, bring your focus back to your breath and notice how your body feels.
10. Repeat this practice for the next five minutes.
11. If you notice your mind is wandering, bring your full attention back to the experience of eating the raisin and noticing any thoughts or emotions that arise.

This mindfulness exercise is a simple way to bring your attention to the present moment and cultivate awareness of your senses. It also helps you to become aware of your mind wandering.

The sound of silence

After my initial practice with meditation, I went on to try the other form of Zen meditation, Rinzai Zen. In this form of meditation, you practice Zazen, which, as mentioned earlier means "just sitting." You attempt to be aware of your thoughts, aware of the sounds, aware of everything. There is no distraction possible since the distraction becomes the center of the meditation. You must think that when practicing this type of meditation, you will immediately become aware when your mind wanders. However, although that can happen, the mind often starts wandering without you being aware of it. Even though your meditation is to be aware and watch your mind drift away, you often fail to do so. You start having random thoughts, only to realize 30 seconds to a minute later that you are lost in them.

What about other meditation traditions?

I then practiced other meditations, including Vipassana and mantra-based meditation. I remember visiting a friend in India and participating in chants in the Ramakrishna tradition. We were gathered in a house in Hubli, a city in India's center. The chanting lasted for about one hour. At the end of the chanting, the swami asked me how it differed from the other meditation types I had practiced. I replied that it felt very similar. I explained that my mind wandered, returned to the chanting, then wandered again. He smiled. I will never forget that smile.

Through meditation training, practitioners learn to focus while simultaneously cultivating equanimity and compassion toward the content of their minds; equanimity is a synonym for mental composure and non-reactivity. As the mind wanders, meditators are instructed to witness and return to their meditation, and in one of the meditations I practiced, the teacher even advised us to label the thought in our mind, for example, "thinking about dinner," before resuming our meditation.

Being the competitive Westerner that I was, I was determined to break through this mind-wandering business. I even designed a small computer program to count how much my mind wandered during meditation, and tried to lower that number every day.

Accepting my thoughts

Still, it does not feel good when your mind wanders during meditation. Even if you are instructed not to judge yourself while meditating, you unconsciously judge your daily meditation by the amount of mind wandering it contains. Was I able to concentrate today, or was my mind all over? When thinking a lot, your meditation also feels shallow. As a meditator, you can become addicted to thinking as a smoker is addicted to smoking. You may feel you have succumbed to the temptation of thinking while meditating. As strange as it seems, it can become a source of stress. Experiencing stress while meditating might seem like an oxymoron, but it does happen. One of the problems with this approach is that mind wandering during meditation is seen as unfavorable.

I later got acquainted with transcendental meditation, another mantra-based meditation. A mantra is a word or sentence you think and repeat internally over and over with great intent. This technique was a breath of fresh air for my meditation practice. Transcendental meditation teaches you that the mind is like a lake and thoughts are like bubbles coming from the bottom of the lake and disrupting the surface. However, this is fine because the thoughts are released in the process. It changed my approach to mind wandering. After I learned this technique, if my mind wandered during meditation, I no longer had to feel bad about it. In this framework, if your mind wanders a lot, it is beneficial as many thoughts bubble up and release their energy. This is the theory, and it is yet unclear if this hypothesis could be backed up by science: does increased mind wandering during meditation lead to more negative thoughts being released and higher levels of equanimity? Whatever the scientific validity of the claim, I know that it helped me stop judging myself on my ability to meditate. I could meditate and relax; if my mind wandered a lot, that was OK.

EXERCISE: POSITIVE MEDITATION

This is what I would like you to try in this section: a meditation where we judge our mind wandering positively and see how that feels. Here's a short breath-counting guided meditation practice that focuses on the acceptance of mind-wandering thoughts:

1. Find a quiet, comfortable place to sit or lie down. Set a timer on your phone for five minutes – or 10 minutes if you feel you are up for it.
2. Close your eyes and take a deep breath through your nose, hold for a few seconds, and exhale slowly through your mouth. This is one breath. Count your breath up to 10 and then restart at 1.
3. Focus on your breath and let go of any thoughts or worries that may be on your mind.
4. If you notice you have lost count, or your mind starts wandering, visualize your mind as a calm lake. Notice any thoughts or emotions that rise to the surface like bubbles.
5. As you observe each thought or emotion, imagine that it is being released from your

mind, allowing its energy to dissipate and dissolve into the air.

6. Take a deep breath and repeat the following phrase to yourself: "This thought has now been released and I am thankful." Resume counting your breath.

7. Continue this process for the next five minutes, focusing on your breath and observing and releasing thoughts as they arise.

Because you are thankful for the thought releasing its energy, you can let it pass without judgment. For example, say you start mind wandering thinking, "Why am I doing this now? I have so much to do. Is this even useful?" Upon noticing the thought, instead of being irritated at yourself for not meditating properly, you are grateful at this thought for releasing its energy.

The science of meditation and mind wandering

Unlike other brain training methods, there is a lot of research on meditation and mind wandering. It is not that meditation is superior to other techniques, simply that there are a lot of people meditating. According to a 2017 US survey, 14% of adults have practiced some form of meditation (Burke et al., 2017). Meditation is an increasingly popular topic in experimental psychology, and there have been numerous studies on meditation and mind wandering that are useful to know about.

We mentioned the endless cycle of focused attention and attentional drift when we discussed what happens in the brain during mind wandering. The person focuses on meditating, but their attention drifts. While they might be aware of having the spontaneous thought, they are unaware their mind has drifted. At some point, they realize their mind is wandering – and we call this moment meta-awareness.

But how does meditation affect mind wandering? I published one of the first experiments on this topic (Brandmeyer & Delorme, 2018). The experiment was simple. My student and I worked with a group of meditators and a computer voice regularly interrupted their meditation, asking them if they were focused on the meditation or if their minds were wandering. This study showed that long-term meditation practitioners had fewer mind-wandering episodes than intermediate practitioners, suggesting that long-term meditation leads to fewer episodes of mind wandering.

However, what if meditation was associated with decreased mind wandering because high-focus participants could self-select to become long-term meditators? Maybe those new to meditation whose minds wander less were more likely to become long-term meditators — whereas others whose minds wander more drop meditation altogether. To show that meditation decreases mind wandering, one must study non-meditators as they learn to meditate, which is what some researchers did. They trained non-meditators to become meditators for two weeks (one hour daily) and observed a reduction in mind-wandering episodes (Mrazek et al., 2013). This was measured using the thought probes we mentioned previously, where participants are interrupted during a task and asked if they are mind wandering. This study showed that not only the number of mind-wandering episodes decrease, but participants also became better at catching themselves mind wandering. Other researchers demonstrated a more significant drop in mind wandering following a few days to a few months of meditation practice (Rahl et al., 2017). The meditation practice varied, although it usually involved some form of mindfulness practice.

* * *

Similar to the Cognitive Behavioral Therapy and Acceptance and Commitment Therapy discussed earlier, meditation allows us to see our thoughts from different perspectives and to reappraise our thoughts through exposure, extinction and reconsolidation.

Through the practice of meditation, upsetting thoughts will come up – this is exposure. In meditation, we cannot simply avoid unpleasant thoughts by occupying ourselves with something else so arguably we might be more exposed to our disturbing thoughts.

Extinction is a concept in psychology indicating the fading and disappearance of thoughts. Meditation is specifically helpful for thoughts that are not too upsetting. Because we realize they pop up repetitively in our minds, we get bored of the thought. Imagine you compete with a colleague at work – you might not want to think about that all the time, especially during your meditation.

Letting go of a thought is different from suppressing it. Letting it go means it naturally loses its grasp on you. As discussed, trying to quash it might make it more potent. Finally, reconsolidation means the thought has been reappraised: the meaning of the thought has changed for you. Before, if the thought crossed your mind, you might have experienced jealousy. After, you experience indifference or even, in the case of positive reappraisal, pleasant emotions.

CHAPTER 14

SEPARATING OURSELVES FROM OUR THOUGHTS

Mind wandering is very similar to breathing – we become aware of it if we focus on it and can regulate it for short periods, but over long periods it is beyond our control. Most interventions affecting mind wandering happen slowly and are based on repeated mental exercises such as meditation.

The Work

All the methods we are looking into have one thing in common: the realization that we are not our thoughts, and especially not our mind-wandering thoughts. This realization is an essential first step. You might have had that experience when a thought pops into your mind about something you might be ashamed of or regret. For example, one of your best friends might have gotten a promotion at work and two spontaneous thoughts compete in your head. Your first thought might be, "Why is she getting a promotion and I am not? I am smarter!" The other one might be, "What is wrong with me? Why can't I be happy for her?" You genuinely love your friend, and you feel ashamed of being jealous. Yet, it shows that you did not control that thought. It came spontaneously even though you might not have wanted it, and wish you did not have it. When we observe our mind's activity, we realize we have little control over some of our thoughts: they arise spontaneously. The realization may also come during activities requiring a high concentration over an extended period. Even if we want to be fully concentrated, we cannot force ourselves to stop mind wandering.

Some mind-training techniques we have seen in the previous chapter do not make the explicit assumption that we are not our thoughts. Other methods I focus on here are more radical in their approach. They start from the premise that we are not our thoughts, that thoughts pass through the stream of consciousness and we decide to believe them.

It is crucial to mention that practices that emphasize we are not our thoughts view the world through a different lens, because if we are not our thoughts, what are we? Often, but not always, the methods derived from Eastern spiritual and meditation traditions, but it is possible to use these techniques without adhering to any underlying worldview.

Can't I just stop believing my thoughts?

Now, it is clearly not enough to state that we are not our thoughts and want to stop having that belief. Below I present several techniques, some based on meditation, some on clinical practices, and some based on spiritual development, to help us to separate from our thoughts.

The "work" of Byron Katie is a form of cognitive restructuring technique to help us to realize we are not our thoughts (Katie, 2002). It is especially useful when we have recurring spontaneous thoughts that we would rather go away. (Byron Katie's method is presented as a spiritual practice, although I believe it can be used in a non-spiritual setting.)

First, we are asked to pick a thought that is upsetting to us. (It is better to start with a moderately upsetting thought.) The method is simple because we are only asked to look deeply into four questions about this thought:

1. "Is it true?"
2. "Can you absolutely know that it's true?"
3. "How do you react when you believe that thought?"
4. "Who would you be without the thought?"

Let's take an example of someone you have strong negative feelings about, like a politician who you believe to be a bad person. Your first answer might be, "Yes, this is true. This person is evil. He did such and such." Upon looking closer at the second question, and whether the statement is absolutely true, you might think, "A lot of people believe in that politician. I do not think he is completely evil. Perhaps he is confused. Perhaps all of his voters are deluded. I surely do not think all of his voters are evil." You might also think that this politician might love their family. The answer to the second question usually is, "No, I cannot know that this is absolutely true." On the third question about how you react when you believe that thought, you might think, "I focus a lot of energy on hating and despising that person. I feel tense when I have that thought." Therefore, you realize that this thought is stressful and detrimental to you. Then on the last question about who you would be without the thought, you think, "I might be more relaxed if I did not have that thought. Maybe if I stopped focusing all my energy on that thought, I might be happier," and that's it! Upon the final realization and with repetitive practice, you recognize that the thought is actually stressful and decide it is not worth continuing to hold it.

EXERCISE: PRACTICING THE "WORK"

- Let's pause and practice. Write something in which you strongly believe on a piece of paper. Now read the first question. Close your eyes and ponder it for at least 30 seconds. Do not hesitate to list all the reasons you believe this is true.
  (1) Is it true?

- When done, open your eyes, and read the second question. Close your eyes again and ponder it for at least 30 seconds.
  (2) Can you absolutely know that it's true?

- When done, move on to the third question.
  (3) How do you react when you believe that thought?

- Ponder the fourth question in the same way.
  (4) Who would you be without the thought?

This introduction is not meant to replace Byron Katie's books and workshops (Katie, 2002), and I invite you to look at that material if you think it might be helpful. Also note that I like to spend

30 seconds on each question, but Byron Katie recommends several minutes. The purpose of spending several minutes with each question is to encourage a deep and honest exploration of one's own thoughts and beliefs. I prefer to spend less time, but then go back to question 1 when I am done with question 4 and repeat the process multiple times.

The Work vs. CBT

Even though the questions above superficially look like cognitive restructuring, they are fundamentally different. Cognitive restructuring is a technique that allows people to identify less with their thoughts and eventually remodel them. However, Byron Katie and other spiritually oriented methods point to the illusion of all thoughts. People practicing these methods are advised to use the technique on both negative and positive thoughts and beliefs. The goal for the practitioners is to realize they are not their thoughts, however disturbing or pleasant. The technique instructs us we are the presence that is beyond thoughts.

There is a famous anecdote about Byron Katie where someone held her at gunpoint in the desert. She did not experience the fear and racing thoughts typical of this situation. Instead, she saw the beauty of the present moment and was curious about what would happen. It is a story worth reading (Katie, 2002).

CHAPTER 15

———————

WORKING WITH BODY SENSATION

Another form of cognitive restructuring involves body sensations – a hugely empowering method for the mastering and understanding of mind wandering. Not only does this help to reduce the occurrence of disturbing thoughts, it also shows us where they reside – in our bodies. Although I had worked with this technique before joining a Zen center in San Diego, my Zen teacher also recommended it to all students. It is one of the ultimate techniques to let go of long-held thoughts when one is ready.

The link between mind and body

This method assumes that our brain is not independent of our body and that many thoughts, in particular mind-wandering thoughts, are anchored in the body. Scientists have indeed shown bidirectional communication between mind/brain and body. For example, studies have shown that stress can affect the immune system and increase the risk of illness (Cohen et al., 2007), while mindfulness meditation has been found to reduce stress and improve immune function (Davidson et al., 2003). Similarly, research has demonstrated that physical activity can improve brain function and reduce the risk of cognitive decline (Erickson et al., 2011) and that mental exercises like cognitive training can improve physical performance (Seeman et al., 2001). These findings suggest that the mind and body are intimately connected, and that interventions targeting either one can have positive effects on the other.

In popular language, we use the expressions "knot in my stomach," "a lump in my throat," or "a heaviness in my heart." William James, the famous psychologist of the early 20th century, proposed that emotions were linked with bodily responses. About 30 years ago, renowned neuroscientist Antonio Damasio refined this model and postulated the somatic marker hypothesis (Damasio, 1994) which states that our body first responds to emotions, which are an interpretation of our body's reaction. For example, if you are scared of something your adrenaline increases, your heart will start beating faster, and you might experience fear. The theory is that the feeling of being scared is the mind's response to what the body experiences, not vice versa.

If, for example, a specific thought makes you feel tense in your back, then your back tension is part of the memory of that thought. The thought lives in your back as much as it lives in your brain. Some studies support this hypothesis. For example, patients who suffer from "pure autonomic failure" have damage to some of their peripheral nerves. They experience no change in blood flow, heart rate or perspiration when confronted with fearful situations. Their body does not respond emotionally. Subjectively, they do not experience emotions as strongly as ordinary individuals, and activity in the brain areas processing emotions is lessened (Critchley, 2002). Because their body does not react to fear, their subjective perception of fear is significantly decreased. So emotions might live in our bodies, and most thoughts are emotionally laden.

One technique leverages this body-mind connection. Physicist and entrepreneur Lester Levenson created the Sedona Method. The story goes that he was given weeks to live after a second coronary heart attack but applied the technique on himself and went on to live for another 42 years. The principle is similar to Byron Katie's "work" and cognitive restructuring. It consists of focusing on a thought – usually negative – and asking oneself three simple questions:

1. "Could I let this feeling go?"
2. "Would I let this feeling go?"
3. "When?"

In this method, you focus not on letting go of the thought but on the associated body sensation. First, you write down a negative thought, for example, "I hate politician X," then you lie down, close your eyes and hold this thought in your mind. How does it make you feel? Does any part of your body feel tense, like your throat, arms or chest? Once you have identified the body's associated tension or sensation, you do not need to remember the thought anymore (although it will likely stay in the back of your mind). Then apply the three questions to the body sensation. This method might sound naive, but it is extremely potent. Let's try the exercise together below, but before that you need to know more about how to work with your sensations.

Working with sensations

Imagine the thought of the evil politician again. You close your eyes, scan your body and realize you feel heaviness in your chest. Then you ask yourself the first question, "Could I let this sensation go?" This question is usually the easiest one. The sensation is typically unpleasant, and the thought of not having it is a no-brainer. You are not asked to actually let go of the sensation, only to consider whether you could let it go on a theoretical level. Imagine one of your possessions. Could you let this possession go if you had to? The answer is probably yes, mainly because it is hypothetical. We are not asking you to let it go yet. It is the same with sensations; we are usually willing to let them go, especially if they are unpleasant.

Letting go of sensations

The second question is about being willing to let go. It is less hypothetical; as you ask yourself this question, you imagine how it would feel without this sensation. This part assumes that you have the choice of keeping or letting go of the sensation. (This obviously only works for sensations associated with thoughts, not something like chronic pain, where you have no choice in the matter.)

Although the sensation is unpleasant, you might experience a void when you imagine yourself without the sensation. Even if unpleasant, it is a part of you. I apologize for the following gory example. It helps to imagine how it might feel. Imagine your legs have become gangrenous due to loss of blood flow and need to be amputated. Would you be willing to remove your own legs? Reason tells you this is the better choice if you want to live, but you might still have second thoughts.

You might not be able to do it and choose to die rather than alter your body image – and endure the associated loss of autonomy. When dealing with body sensations, it can feel similar. Letting go of a sensation means you have to let go of part of yourself, which might be hard to accept. This sensation, although unpleasant, makes you feel alive. Your mind might convince you that it is not that bad and that you can live with the sensation. You might intellectually want the sensation to go away but not be willing to let it go.

The last question is practical. It asks if you can let it go now, as you might want to hold on to it a little longer. Again, it is more important to be true to yourself than to force yourself to let go of a sensation you are not truly ready to let go of.

The reasoning behind these questions is to have you look deeply at the link between the sensations and the negative thoughts you are experiencing. The three questions could be summarized in a single "look deeply at the sensations this thought triggers," but being asked to make a choice increases your willingness to look at these sensations (which are often uncomfortable). Also, it is fine to answer "no" to any of the three questions. It is more important to be true to oneself than to let go of the sensation.

Letting go of the thought

Once you finish question three, return to the first one. Using this technique, I have also observed that there may be a delayed effect. You could have worked really hard on your thoughts and sensations, and apparently nothing happened. Yet, a few hours later, you feel deeply peaceful.

This method can have a significant effect on mind wandering. Spontaneous negative thoughts you experience during mind wandering are often core worries or blockage – the rumination discussed earlier. By using the Sedona Method to let them go, you stop experiencing them when you mind wander. You may mind wander less, and the content of thoughts when you are mind wandering will change for the better – which we know increases happiness and satisfaction.

The Sedona Method works like magic because the thought is gone once you truly let go of the sensation. The coupling between your brain and your body kept this thought alive – the thought is living in your body as much as your mind. When you succeed in decoupling them – by realizing you do not want the unpleasant sensation – the thought vanishes. This is true even if you have forgotten the thought while focusing on the sensation.

EXERCISE: THE SEDONA METHOD

Here is an example of practicing the Sedona Method.

1. Take a piece of paper. Write down a spontaneous thought that is causing you stress, discomfort, anxiety, or anger.
2. Lie down on your back and set a timer for five minutes.
3. Feel within your body any sensation or tension as you think the upsetting thought. Do this for about one minute. This sensation will usually be unpleasant.
4. Ask yourself the following questions:
   a) Would I be willing to let go of this sensation right now? Spend 30 seconds to see how it feels to want this sensation to go.
   b) Could I let it go? Spend 30 seconds to see how it feels to let it go.
   c) When could I let it go? Spend 30 seconds to see how it feels to let it go now.
5. Repeat these questions for five minutes, focusing on the sensation.

When done, you should notice how the original upsetting thought has lost some grip on you.

> This is just a basic example of the Sedona Method (Dwoskin, 2003), and there are many variations and approaches to the practice.

After a few repetitions of the three questions, you might be genuinely willing to let go of the sensation, but you are not done. The sensation will come back and, with it, the negative thought. You have to do it repeatedly until you retrain your brain and body. I like to imagine the negative sensation living in the body like a bear living in his den. You will chase the bear off once, and it will return to its home. Only after you chase it over and over will the bear finally go away.

Over decades of conditioning, the reaction and association between thought and body sensation have become second nature. This might correspond to neural and hormonal pathways at the brain and body levels. You must work for some time to decrease the strength of the neural pathway and recondition your brain and body. Note that a lot of meditation techniques work similarly. As you meditate, scientists have shown that you become more aware of body sensations, which raise awareness of their association with unpleasant thoughts. Again, this is relevant for mind wandering, since, as you let go of your negative thoughts, your mind wandering will not be as disturbing.

Working with difficult thoughts

A spontaneous thought about your hatred for a politician is a relatively easy thought to let go of. However, looking into other topics, such as cravings, chronic anger or relationship issues can become challenging and overwhelming. You see the negative sensation, but are in no way willing or capable of letting it go at first. It is part of yourself. It is important to go slow because letting go too fast could lead to serious side effects where your body and mind react negatively, bringing on extreme anxiety or panic attacks. When you experience these, you might be tempted to fall back into your old habits, which, although uncomfortable and ruining your life, are judged better by your mind than the extreme anxiety and reaction encountered when trying to let them go. It will require tremendous courage and perseverance to move forward, but know that this state is transient. Go slower and work with less challenging thoughts to avoid the dramatic side effects. If you persevere, you can return to a new normal where the thoughts are mostly gone. You are a little bit happier, and ready for the much work that still needs to be done.

CHAPTER 16

THE HAPPINESS TECHNIQUE

The main goal for taming our wandering minds is to reach greater happiness or peace. Doing the exercises in this book, it is sometimes difficult to see if we are on the right path. Even if we are ready for the change, our bodies and mind will take extreme measures to prevent that. We might experience anxiety attacks and intense psychosomatic body pain because our very being resists change. However, once we finally break through, after a few weeks or months, we often resume our former identity as if little had happened. We are probably a little happier because we no longer have the former limitations destroying our lives, but we still have problems in our lives.

The happiness meter

One way to know if we are on the right path is "the happiness meter." If you have worked hard to gain a greater control over your wandering mind, it might have been extremely taxing physically and mentally, and you might ask yourself if it was worth it. How can you know you are better off? The happiness meter measures if you are getting happier. If the answer is positive, you are on the right track. And knowing if we are happy might simply be looking at how happy and at peace we feel now. Or it might be to look at our job and relationship satisfaction and compare it to what it was before. I used to be a meditation and mind-wandering researcher because I thought it was the key to happiness. However, I have now cut the middle-man and consider myself a researcher on happiness.

The Happiness Method – and its associated happiness meter – is not a structured method. It consists of periodically (say a couple of times a month) assessing if whatever we are doing makes us happier. It is used in conjunction with the other techniques to judge if they work for us. The happiness meter is good for getting yourself out of negative thoughts, and could also be used as a general principle for life. Were you right to change jobs? Focus on comparing on how happy you were before and how happy you are now: use the happiness meter. Are you in a better relationship than before? The happiness meter will tell you. Was it a good idea to move? The happiness meter. Do you see your family and friends enough? The happiness meter.

The advantage of the happiness meter is that it weighs everything in your life. Say you enjoy hanging out with your family but they drive you crazy. Yet you feel guilty and spend more time with them than you want. Does the feeling of being a good family member bring you more happiness than your desire to avoid your family would? By focusing on what makes you happy, you can make the best choice for you. And this goes not only for everyday concerns but also for your overall peace of mind and mind wandering. You can use your subjective meters of internal peace and happiness to know if you are on the right path.

I want to give a simple, almost ridiculous, example. I used to restrain myself to only reading books that helped me grow – self-help books and scientific books. However, these generated a lot of mind wandering. I learned by reading these books, but it was not always fun; my mind often went off on tangents. I used to believe that life was too short to spend it reading fiction. Now, I have let go of that belief. I read fiction books I like, and I enjoy them thoroughly. My mind no longer wanders when I read, and it is a joy to read. My happiness meter shows me it was a good choice. I am happier for having made that choice, and I am more pleasant for being slightly happier, so the happiness of people around me is also affected. It actually also helped me to regulate my emotions, because I have a "good place" that nobody can take away from me, where I am absorbed and at peace. For other people, it might be running, or some other activity. You will notice, however, that your good place is where you do not mind wander too much – remember the Harvard happiness study? So decreasing your mind wandering might be as simple as changing the activities to somewhere you mind wander less and are more happy.

EXERCISE: THE HAPPINESS METHOD

Here is an example of practicing the Happiness Method:

1. Take a piece of paper and create two columns on it: "Positives" and "Negatives," and fill in the good and bad aspects of your life currently. This could be your job, your relationships, your health.
2. Now, cast your mind back one year and look at the listed situations in comparison to one year previously – perhaps you are earning more than the previous year and getting on better with your partner (positives) but you might have a difficult relationship with a new boss.

3. Reflect on what changes you have made in your life that have made situations better or worse, and vow to continue on a path that increases happiness in your life.

This exercise is based on introspection – how happy you are feeling. There is no right or wrong way to judge your life.

What is happiness?

It is essential to understand what I mean by happiness. Happiness is not short-term pleasure. By happiness, I mean long-lasting peace. Yes, of course, eating that ice cream right now will bring you happiness in the present moment, but will it bring you lasting peace?

And we do what we can, depending on who and where we are. Our paths may even cross. For me, letting go of reading "useful books" brought me happiness. For someone else, it might be the opposite. There is no right or wrong. We sometimes need to make errors and end up on the wrong path to find our way to peace. For example, if we believe a career change would make us happier, we should probably do it, even if we return to our original job because we realize it was not for us. After we return, we know ourselves better and are more content doing what we do because we do not fantasize about doing something else.

* * *

Before returning to how to curb mind wandering and be happier, let me finish by pointing out a few non-evident lifestyles that research has shown to make people happier. Increased income makes people happier, as you might expect, but not by a significant amount unless they live in poverty. Living in a community you trust makes you happier (Lopez-Ruiz et al., 2021). Being self-confident in what you do and working less makes people happier. Being surrounded by nature also makes people happier. In other words, if you are good at your job and living in a rural community you trust and love, you might think twice about changing careers for a demanding higher-paid job, forcing you to move to the center of a big city you do not know. And I would argue – although this is a hypothesis at this point – that a lot has to do with the content of mind wandering. If you are in a community you trust, your mind does not need to wander about what others will do to you. If you have a stress-free job, your mind might not often wander about what could go wrong, making you unhappy.

CHAPTER 17

OTHER TECHNIQUES FOR OVERWHELMING THOUGHTS

When mind-wandering thoughts are overwhelming, such as around a trauma you might have experienced, working with any of the techniques mentioned above is difficult. The techniques above could lead to extreme anxiety we are not ready for. I recommend the methods below for dealing with overwhelming thoughts such as rumination, obsessive thinking and mind wandering. These techniques involve splitting your attention. By forcing you to focus on two things at the same time, the upsetting thought and the behavioral practice, the upsetting thought cannot grab your whole attention and becomes less overwhelming and more acceptable.

EMDR

The first technique is called Eye Movement Desensitization and Reprocessing (EMDR), and is usually practiced with a therapist. It is a mouthful but much more straightforward than it sounds: simply look at an object. The method consists of recalling traumatic thoughts and practicing eye movement patterns at the same time, where the therapist might ask you to follow a pen with your eyes. You are not hypnotized. As you follow the pen while recalling traumatic experiences, the traumatic thought becomes more acceptable. One hypothesis is that our capacity for working memory is limited. By asking patients to perform two tasks simultaneously, the traumatic image and sensations become less overwhelming and easier to handle for the patient. In time, they accept the sensations as bearable, and their relationship with the trauma changes. The EMDR technique is often used to treat post-traumatic stress disorder (PTSD).

The EFT tapping method

The Emotional Freedom Technique, or EFT for short, is another method to help process overwhelming thoughts and feelings. It involves tapping one or two fingers on different body parts a few times per second. Tapping is often performed over areas with high nerve densities, such as around the eyes or under the nose. It is unknown why this technique works, but the mechanism might be similar to EMDR. The patient's attention is split, and they cannot fully engage in the memory of their current or past trauma. EMDR and EFT involve voluntary muscular actions of the eyes or the fingers. EFT adds sensations with the tip of fingers touching part of the body.

EXERCISE: EFT TECHNIQUE VARIANT

Here's a five-minute exercise using a variation of the Emotional Freedom Technique (EFT) finger tapping method:

1. Identify a specific fear, anxiety, trauma or negative emotion that you would like to work on.
2. On a scale of 0 to 10, rate the intensity of the negative thought.
3. Set a timer for five minutes.
4. Use your first two fingers to tap around your eye as you are trying to relive the negative emotion as shown in the figure below.
5. You should tap about three to four times per second, and in all cases more than once per second. As you tap, move your fingers

around your eyes. Note that it is normal for the intensity to fluctuate during the tapping process. The important thing is to stay focused and persistent in tapping, and the intensity will eventually reduce.

6. Let your mind dwell on the fear, anxiety, trauma or negative emotion that you have identified prior to the tapping process.

7. Do so for five minutes until your timer runs out, or until the emotion decreases in intensity.

8. After five minutes, rate again the intensity of the negative thought. It should have decreased in intensity.

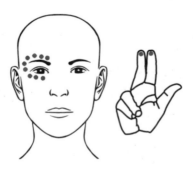

Other points people tap when using the EFT method are under the nose, the chin, the collarbone, and under the arm – usually locations on the body identified as acupuncture points.

Biofeedback

Another technique that has proven efficient at handling trauma is biofeedback, in particular, the method developed by Heartmath. A Heartmath device measures heart rate and heart variability (the change in timing between heartbeats). If the timing is very regular, it is usually not a good sign and could indicate a risk of a heart attack. By contrast, if the heart rate is irregular (within limits), it means involvement of the parasympathetic system, which is the relaxation system of the body. The Heartmath device works by emitting beeps of different tones based on heart rate variability. The user aims to produce the beeps associated with high heart rate variability and relaxation. In Heartmath therapy, such as the one I have published with my psychiatrist friend and colleague Stephanie Hahusseau, patients may be asked to recall unpleasant thoughts while performing biofeedback (Hahusseau et al., 2020). By remembering unpleasant thoughts and forcing their body to remain relaxed, they come to accept these thoughts.

Other methods

Interestingly, running has also been shown to increase emotional well-being and mental health. It might involve similar mechanisms as EFT such as rhythmic sensations while rehearsing negative thoughts. So if your mind wanders on negative thoughts, you might as well do it while running.

Many other effective behavioral methods exist to process unpleasant, spontaneous pervasive thoughts occurring during mind wandering. For example, hypnotic suggestions can decrease pain and emotional stress associated with emotional thoughts. Psychedelic therapy has also been shown to reset the mind and allow patients to move past trauma, depression and addiction. Lucid dreaming can also help people face and overcome their fear. As referenced above, sports and reconnecting to one's body sensations and their association with thoughts have also proven efficacious in treating mental illnesses like depression.

Willingness, willingness, willingness

Letting go of thoughts and changing oneself to become a better person is very much about the willingness to make it happen. This statement might sound cruel to those suffering, but I also want to acknowledge the difficulty of making change. In extreme cases, you need to be ready to "die" to move forward. And you will – a part of you will die for the better.

The willingness factor

One of my dear friends was diagnosed with bipolar disorder and was hospitalized in his twenties. Bipolar individuals oscillate between extremely depressed and hyperactive. One of my friend's problems is that he enjoyed the highs. He felt genuinely alive at these moments, even though they wreaked havoc on his professional and personal life. He was highly sociable in these episodes, hyper confident and often bypassed usual interpersonal boundaries. But he could not sleep more than a couple of hours a night. The episode would usually culminate. He would do something really dumb and finally come down. He would then become depressed for a couple of months. In his case, I believe he could not completely let go of wanting to experience these states, which is why he relapsed every couple of years. Once he chose to experience the high, it was almost impossible for him to return until he hit rock bottom. Over the years, he has learned that the highs come at a price and has accepted not to experience them.

The principle is the same for disturbing thoughts during mind wandering. If one genuinely wants to stop having these thoughts, I believe one can use the above-mentioned techniques and it will work. It might take some deep looking the first time, but over time, it should become automatic. As soon as a negative thought linked with depression creeps in, the mind automatically says, "No, thank you!" This only works for strong and genuine intent, though, where people feel their life in balance. Note that this by no means suggests that those who are depressed want to experience that state. It only implies that they are caught in a vicious circle and can escape it with great courage, effort, and perseverance.

How to apply mind-changing techniques

The techniques presented in the previous sections are not magic. Like many people, I only turn to these techniques in periods of crisis, when I believe I need to do something about my negative thinking and intrusive mind wandering. If you have picked up this book, I believe you are prepared to do the work. These methods have brought me and others comfort because I feel that if I ever fall into a negative vicious circle, I have the tools to retrain my mind for the better.

Miracle cures may work

Often when one is ready to change, the method or practice is just a catalyst. The technique is not the cause of the change. Only when someone becomes hopeless are they ready to take the leap of faith.

Spiritual teacher Eckhart Tolle is an excellent example of that. He had to come close to suicide before he could let go of his former unhappy, self-destructing thoughts and identity. He describes the story of being in the subway and observing a schizophrenic individual vocalizing negative thoughts. This episode made him realize he was experiencing the same madness and constant mind wandering about negative thinking. He later had an epiphany, was able to let go of his negative thoughts and old identity, and lived for more than a year on a park bench in bliss (Tolle, 2004).

* * *

We have seen that while occasional mind wandering is normal, excessive and persistent mind wandering can cause a range of negative effects, from decreased productivity and reduced attention span to increased stress and anxiety. Fortunately, various techniques can help individuals address and reduce their mind-wandering tendencies. These techniques help us develop a greater awareness of our thoughts, emotions and bodily sensations. By learning to recognize when our minds start to wander, we can take steps to redirect their attention and focus on the present moment. This increased self-awareness can also help us identify the triggers that cause our minds to wander, such as stress, boredom or fatigue, and develop strategies to manage these triggers more effectively. Moreover, these therapies help us develop skills that promote attentional control and improve our ability to sustain focus on a task. By learning to manage our mind-wandering tendencies, we can make better use of our time, reduce stress and anxiety levels, and enhance our overall well-being.

CONCLUSION

Willpower alone is not enough to make lasting change, but it is essential for taking action and sticking with a practice over time. It is important to have a clear understanding of why you want to use these techniques, and to set achievable goals for yourself. In order to reap the benefits of these techniques, it is necessary to be committed to the process and to practice regularly. Without this commitment, it can be easy to fall back into old patterns and habits, and the benefits will not be sustained.

I am not my wandering mind

The mind is a powerful tool that can significantly influence our happiness and life fulfillment: our spontaneous thoughts and beliefs shape our perception of the world around us, which in turn affects our emotions, behavior and overall well-being. We also saw how we can train our minds to focus on the good in our lives and develop a greater sense of contentment and fulfillment.

We have seen how the power of science to learn more about the brain and how it works has enormous potential for the future. With new technologies and techniques, researchers are able to uncover the underlying mechanisms of brain function and identify ways to optimize cognitive performance and mental health. By understanding how the brain processes information and responds to different stimuli, we can develop more effective interventions and treatments to curb our negative mind wandering.

Meditation, cognitive restructuring and body centered methods cannot be overstated in their potential to seriously impact how we experience our world. By practicing mindfulness and focusing on the present moment, we can develop greater self-awareness and learn to respond to situations in a more adaptive way. By tuning into our body sensations and learning to identify and regulate our emotions, we can reduce stress, anxiety and other negative emotions that impact our quality of life. These techniques offer a powerful and effective means of promoting mental and emotional well-being.

* * *

I would like to end with a quote by spiritual teacher Rupert Spira. Spira is a contemporary spiritual teacher and author based in the UK, teaching that thoughts are not ultimately real, but like images that appear and disappear in the screen of awareness. He emphasizes that thoughts are simply mental events that arise and pass away within the field of consciousness, and do not have an inherent reality of their own. According to Spira, the mistake that many people make is in identifying with their thoughts, and believing that they are the thinker of their thoughts. He teaches that the true self is the consciousness in which thoughts, and all other mental and sensory experiences, appear and disappear. Spira suggests that by recognizing the impermanence and insubstantiality of thoughts, we can free ourselves from the limitations and suffering that arise from identifying with them. He encourages a practice of observing thoughts without judgment or attachment, and instead resting in the awareness that underlies all mental activity.

"Thoughts are just like birds flying across the sky of awareness. They appear and disappear in the boundless space of your being."

Rupert Spira
(Spira, 2021)

ACKNOWLEDGEMENTS

I dedicate this book to my wife, Rosanna, and my two daughters, Chiara and Leili, who made me the person I am today; to my parents, my brothers, and their family with my many nieces and nephews, who bring the word family to life in all its glory; to friends Bruno, Patrice, Rael, Tim, and Stephanie, who know my imperfections and love me, nonetheless; to my academic coworkers and mentors, Helané, Dean, Gary, Marilyn, Scott, Dung, Cyril, Robert, Fiorenzo, Ramon, Ravindra, Rufin, Michele, and Simon, who have inspired me and taught me most of what I know; to my former Ph.D. students, Tracy, Cedric, Claire, and Romain, who entrusted me with such an important part of their life; to my editors, Beth and Jo, and proofreaders Kate and Sitara, for their patience, and dedication to bring this book to life, and to Jonathan, one of my heroes, for writing the foreword; finally, to my spiritual teachers Joko Beck and Ruppert Spira, for their guidance in dealing with my wandering mind.

REFERENCES

Anderson, T., et al. (2021). "The metronome response task for measuring mind wandering: Replication attempt and extension of three studies by Seli et al." *Atten Percept Psychophys* 83(1): 315–330.

Andrews-Hanna, J. R., et al. (2010). "Evidence for the default network's role in spontaneous cognition." *J Neurophysiol* 104(1): 322–335.

Anonymous (2018). *The Cloud of Unknowing*, Boston: Shambhala Pocket Edition.

Arabaci, G. and Parris B. A. (2018). "Probe-caught spontaneous and deliberate mind wandering in relation to self-reported inattentive, hyperactive and impulsive traits in adults." *Sci Rep* 8(1): 4113.

Baird, B., et al. (2012). "Inspired by distraction: mind wandering facilitates creative incubation." *Psychol Sci* 23(10): 1117–1122.

Braboszcz, C., et al. (2017). "Increased gamma brainwave amplitude compared to control in three different meditation traditions." *PLoS ONE* 12(1): e0170647.

Braboszcz, C. and A. Delorme (2011). "Lost in thoughts: neural markers of low alertness during mind wandering." *Neuroimage* 54(4): 3040–3047.

Brandmeyer, T. and A. Delorme (2018). "Reduced mind wandering in experienced meditators and associated EEG correlates." *Exp Brain Res* 236(9): 2519–2528.

Brewer, J. A., et al. (2011). "Meditation experience is associated with differences in default mode network activity and connectivity." *Proc Natl Acad Sci USA* 108(50): 20254–20259.

Bremer, B., et al. (2022). "Mindfulness meditation increases default mode, salience, and central executive network connectivity." *Sci Rep* 12(1): 13219.

Burke, A., et al. (2017). "Prevalence and patterns of use of mantra, mindfulness and spiritual meditation among adults in the United States." *BMC Complement Altern Med* 17(1): 316.

Cahn, B. R., et al. (2010). "Occipital gamma activation during Vipassana meditation." *Cogn Process* 11(1): 39–56.

Chen, E. (1979). "Twins reared apart: A living lab." *The New York Times*, Dec. 9, 1979.

Cohen, S., et al. (2007). "Psychological stress and disease." *JAMA* 298(14): 1685–1687.

Collischon, M. (2019). "The returns to personality traits across the wage distribution." ... *Labour* 12165.

Critchley, H. D., Mathias, C. J., & Dolan, R. J. (2002). "Fear conditioning in humans: the influence of awareness and autonomic arousal on functional neuroanatomy." Neuron, 33(4), 653–663.

Damasio, A. (1994). *Descartes' Error: Emotion, Reason, and the Human Brain*. G.P. Putnam's Sons. https://ahandfulofleaves.files.wordpress.com/2013/07/descartes-error_antonio-damasio.pdf

Davidson, R. J., et al. (2003). "Alterations in brain and immune function produced by mindfulness meditation." *Psychosom Med* 65(4): 564–570.

Descartes, R. (1641). *Meditations on First Philosophy*. Hackett Publishing Company; 3rd edition (1993).

REFERENCES

Dwoskin, H. (2003). *The Sedona Method: Your Key to Lasting Happiness, Success, Peace and Emotional Well-being.* Sedona Press.

Erickson, K. I., et al. (2011). "Exercise training increases size of hippocampus and improves memory." *Proc Natl Acad Sci USA* 108(7): 3017–3022.

Freud, S. (1908, 1962). "*Creative writers and daydreaming.*" in J Strachey (Ed.) *The Standard Edition of The Complete Psychological Works of Sigmund Freud.* London: Hogarth, Vol. IX.

Fultz, N. E., et al. (2019). "Coupled electrophysiological, hemodynamic, and cerebrospinal fluid oscillations in human sleep." *Science* 366(6465): 628–631.

Gazzaniga, M., et al. (1998). *Cognitive Neuroscience: The Biology of the Mind,* W. W. Norton & Company.

Gillihan, S.J. (2018) "Retrain Your Brain: Cognitive Behavioral Therapy in 7 Weeks: A Workbook for Managing Depression and Anxiety". Althea Press.

Groot, J. M., et al. (2022). "Catching wandering minds with tapping fingers: neural and behavioral insights into task-unrelated cognition." *Cereb Cortex* 32(20): 4447–4463.

Hahusseau, S., et al. (2020). "Heart rate variability biofeedback intero-nociceptive emotion exposure therapy for adverse childhood experiences." *F1000Res* 9: 326.

Harris, R., Hayes, S.C. (2019) "ACT Made Simple: An Easy-To-Read Primer on Acceptance and Commitment Therapy". 2nd Second Edition. New Harbinger.

Irrmischer, M., et al. (2018). "Strong long-range temporal correlations of beta/gamma oscillations are associated with poor sustained visual attention performance." *Eur J Neurosci* 48(8): 2674–2683.

Jackson, J. D. and Balota, D. A. (2012). "Mind-wandering in younger and older adults: converging evidence from the Sustained Attention to Response Task and reading for comprehension." *Psychol Aging* 27(1): 106–119.

James, W. (1890). *The Principles of Psychology*. Henry Holt and Co.

Jana, S. and Aron, A. R. (2022). "Mind wandering impedes response inhibition by affecting the triggering of the inhibitory process." *Psychol Sci* 33(7): 1068–1085.

Jang, K. L., et al. (1996). "Heritability of the big five personality dimensions and their facets: a twin study." *J Pers* 64(3): 577–591.

Joormann, J., et al. (2012). "Neural correlates of automatic mood regulation in girls at high risk for depression." *Journal of Abnormal Psychology* 121: 61–72.

Kabat-Zinn, J. (1990). *Full Catastrophe Living: Using the wisdom of your body and mind to face stress, pain, and illness*. New York, Deltacorte.

Kane, M. J., et al. (2007). "For whom the mind wanders, and when: an experience-sampling study of working memory and executive control in daily life." *Psychol Sci* 18(7): 614–621.

Katie, B. (2002). *Loving What Is: Four questions that can change your life*. Crown Archetype.

Kaur, J., et al. (2021). "Waste clearance in the brain." *Front Neuroanat* 15: 665803.

Keulers, E. H. H. and Jonkman, L. M. (2019). "Mind wandering in children: Examining task-unrelated thoughts in computerized tasks and a classroom lesson, and the association with different executive functions." *J Exp Child Psychol* 179: 276–290.

Killingsworth, M. A. and Gilbert, D. T. (2010). "A wandering mind is an unhappy mind." *Science* 330(6006): 932.

Lazar, S. W., et al. (2005). "Meditation experience is associated with increased cortical thickness." *Neuroreport* 16(17): 1893–1897.

López-Ruiz, V. R., et al. (2021). "The relationship between happiness and quality of life: A model for Spanish society." *PLoS ONE* 16(11): e0259528.

Marcusson-Clavertz, D., West, M., Kjell, O.N.E. and Somer, E. A daily diary study on maladaptive daydreaming, mind wandering, and sleep disturbances: Examining within-person and between-persons relations. PLoS One. 2019 Nov 27;14(11):e0225529. doi: 10.1371/journal. pone.0225529. PMID: 31774836; PMCID: PMC6880993.

Mirsal, H., et al. (2004). "Childhood trauma in alcoholics." *Alcohol Alcoholism* 39(2): 126–129.

Mooneyham, B. W. and Schooler, J. W. (2013). "The costs and benefits of mind-wandering: A review." *Can J Exp Psychol* 67(1): 11–18.

Mrazek, M. D., et al. (2013). "Mindfulness training improves working memory capacity and GRE performance while reducing mind wandering." *Psychol Sci* 24(5): 776–781.

Muller, M., et al. (2021). "Mind-wandering mediates the associations between neuroticism and conscientiousness, and tendencies towards smartphone use disorder." *Front Psychol* 12: 661541.

Noftle, E. E. and Gust, C. J. (2019). "Age differences across adulthood in interpretations of situations and situation-behaviour contingencies for Big Five states." *Eur J Pers* 33(3): 279–297.

Playfair, G. L. (1999). "Identical twins and telepathy." *Journal of the Society of Psychical Research* 63: 854.

Rahl, H. A., et al. (2017). "Brief mindfulness meditation training reduces mind wandering: The critical role of acceptance." *Emotion* 17(2): 224–230.

Raichle, M. E., et al. (2001). "A default mode of brain function." *Proc Natl Acad Sci USA* 98(2): 676–682.

Roberts, W. A. and Feeney, M. C. (2009). "The comparative study of mental time travel." *Trends Cogn Sci* 13(6): 271–277.

Rusting, C. L. and Larsen, R. J. (1998). "Personality and cognitive processing of affective information." *Personality and Social Psychology Bulletin* 24(2): 200–213.

Seeman, T. E., et al. (2001). "Allostatic load as a marker of cumulative biological risk: MacArthur studies of successful aging." *Proc Natl Acad Sci USA* 98(8): 4770–4775.

Seli, P., et al. (2018). "The family resemblances framework for mind-wandering remains well clad." *Trends Cogn Sci* 22(11): 959–961

Sington, D. (2007). *In the Shadow of the Moon*. Produced by Film4, Passion Pictures and Discovery Films.

Smallwood, J. and Schooler, J. W. (2015). "The science of mind wandering: Empirically navigating the stream of consciousness." *Annu Rev Psychol* 66: 487–518.

Smallwood, J., et al. (2009). "When is your head at? An exploration of the factors associated with the temporal focus of the wandering mind." *Conscious Cogn* 18(1): 118–125.

Smith, G. K., et al. (2018). "Mind-wandering rates fluctuate across the day: Evidence from an experience-sampling study." *Cogn Res Princ Implic* 3(1): 54.

Soemer, A. and Schiefele, A. (2020). "Working memory capacity and (in)voluntary mind wandering." *Psychon Bull Rev* 27(4): 758–767.

Spira, R. (2021). "What is awareness?" Rupert Spira's YouTube Channel.

Targ, R. and Puthoff, H. (1974). "Information transmission under conditions of sensory shielding." *Nature* 251: 602–607.

Taylor, V. A., et al. (2013). "Impact of meditation training on the default mode network during a restful state." *Soc Cogn Affect Neurosci* 8(1): 4–14.

Tolle E., (2004). *The power of the Now*. New World Library.

Tomasino B, Chiesa A, Fabbro F. Disentangling the neural mechanisms involved in Hinduism- and Buddhism-related meditations. Brain Cogn. 2014 Oct;90:32-40. doi: 10.1016/j.bandc.2014.03.013. Epub 2014 Jun 27. PMID: 24975229.

VanRullen, R. (2016). "Perceptual cycles." *Trends Cogn Sci* 20(10): 723–735.

Wahbeh, H., et al. (2022). "Exploring personal development workshops' effect on well-being and interconnectedness." *J Integ Complemen Med*, *28*(1), 87–95.

Watson, D., et al. (1994). "Structures of personality and their relevance to psychopathology." *J Abnorm Psychol* 103(1): 18–31.

Xie, L., et al. (2013). "Sleep drives metabolite clearance from the adult brain." *Science* 342(6156): 373–377.

Zanesco, A. P. (2020). "Quantifying streams of thought during cognitive task performance using sequence analysis." *Behav Res Methods* 52(6): 2417–2437.

Zani, A., et al. (2020). "Electroencephalogram (EEG) alpha power as a marker of visuospatial attention orienting and suppression in normoxia and hypoxia. An exploratory study." *Brain Sci* 10(3): 140.

FURTHER RESOURCES

Books

Kabat-Zinn, J. (1990). *Full Catastrophe Living: Using the wisdom of your body and mind to face stress, pain, and illness.* New York, Deltacorte.

Dwoskin, H. (2003). *The Sedona Method: your key to lasting happiness, success, peace and emotional well-being.* Sedona Press.

Katie, B. (2002). *Loving What Is: Four questions that can change your life.* Crown Archetype. Cohen, S., et al. (2007). "Psychological stress and disease." JAMA 298(14): 1685–1687.

YouTube Channels

IONS YouTube Channel
www.youtube.com/@InstituteofNoeticSciences

Rupert Spira's YouTube Channel
www.youtube.com/@rupertspira